T0135964

Proceedings

of the

International Beilstein Symposium

on

FUNCTIONAL NANOSCIENCE

May 17th – 21st, 2010

Bozen, Italy

Edited by Martin G. Hicks and Carsten Kettner

BEILSTEIN-INSTITUT ZUR FÖRDERUNG DER CHEMISCHEN WISSENSCHAFTEN

Trakehner Str. 7 – 9
60487 Frankfurt
Germany

Telephone:	+49 (0)69 7167 3211	**E-Mail:**	info@beilstein-institut.de
Fax:	+49 (0)69 7167 3219	**Web-Page:**	www.beilstein-institut.de

IMPRESSUM

Functional Nanoscience, Martin G. Hicks and Carsten Kettner (Eds.), Proceedings of the Beilstein-Institut Symposium, May 17th – 21st 2010, Bozen, Italy.

Bibliographic information published by the *Deutsche Bibliothek*. The *Deutsche Bibliothek* lists this publication in the *Deutsche Nationalbibliografie*; detailed bibliographic data are available in the Internet at http://dnb.ddb.de.

ISBN 978-3-8325-2901-7

Layout by:	Hübner Electronic Publishing GmbH	Printed by	Logos Verlag Berlin GmbH
	Steinheimer Straße 22a		Comeniushof, Gubener Str. 47
	65343 Eltville		10243 Berlin
	www.huebner-ep.de		www.logos-verlag.de

Cover Illustration by: Bosse und Meinhard
Kaiserstraße 34
53113 Bonn
www.bosse-meinhard.de

PREFACE

The Beilstein Symposia address contemporary issues in the chemical and related sciences by employing an interdisciplinary approach. Scientists from a wide range of areas – often outside chemistry – are invited to present aspects of their work for discussion with the aim of not only advancing science, but also, furthering interdisciplinary communication.

Feynman's engaging title for his 1959 lecture, "There's plenty of room at the bottom" is as valid now as it was when he gave it. He presented a vision of a scientific world beyond a few billionths of a meter that was at that time far away of any technological feasibilities and applications. However, it opened the minds towards the creation of new scientific disciplines that are now called nanoscience and nanotechnology. The "nano" prefix not only refers to the extremely small but also stands for the integration of traditional physics, chemistry, biology and engineering disciplines to form an interdisciplinary science which has far-reaching consequences for science, the environment and society.

Scientific research is about gaining knowledge of a system, which technology can then use for developing practical applications. In the nanoscale dimension, there are unrivalled possibilities for the development of functional objects and techniques in areas ranging from nanoelectronics, nanoscale sensors and novel data storage devices to novel materials and coatings, cosmetics, fuel cells, catalysts, to pharmaceuticals and medical implants. The properties and phenomena that these objects exhibit occur precisely because they are extremely small, existing in an environment where the laws of physics operate in unfamiliar ways. Today, the full ramifications of many experimental achievements are not always apparent and how many of these will result in applications in the future is unclear – the potential is perhaps only limited by our own imagination.

One of the main challenges of nanoscience and technology over the next decades is to achieve precise positional control of material at the nanoscale allowing, for example, the fabrication and manipulation of single molecules. Top-down approaches, such as lithography or bottom-up, as in biological systems, combined with imaging and manipulation techniques such as STM and optical tweezers are providing scientists with insights into the behavior and control of matter at the nanoscale.

In nature, we find complex, highly efficient and highly optimized systems such as biological cells which demonstrate how matter and energy can be controlled on the nanoscale. A higher degree of understanding of how biological systems are organized and function will not only increase our knowledge of living things but will find applications in other branches of nanoscience. This will not only enhance our ability to manufacture functional materials, but also holds promise to find solutions to more general problems in, for example, the areas of energy and health.

This symposium on Functional Nanoscience brought together experts in the field to present and discuss new results and approaches including the following aspects of nanoscience and nanotechnology, i. e. self-organization and self-assembly, molecular motors and transport, self-replicating biomimetic systems, quantum effects, molecular magnets, imaging and manipulation of molecules at the atomic scale / single molecule reactions.

 **Beilstein-Institut**

We would like to thank particularly the authors who provided us with written versions of the papers that they presented. Special thanks go to all those involved with the preparation and organization of the symposium, to the chairmen who piloted us successfully through the sessions and to the speakers and participants for their contribution in making this symposium a success.

Frankfurt/Main, June 2011

Martin G. Hicks
Carsten Kettner

Cover Image, courtesy of Michael Huth, University of Frankfurt/Main

Scanning electron microscopy image of a Pt-C composite structure created by focused electron beam induced deposition (FEBID) using trimethyl-Pt-cyclopentadienyl-methyl as precursor gas. The deposit resulted from a repeated electron beam raster process with 10 nm pitch and 100 micro-seconds dwell time on each point colored in black of a black-and-white bitmap defining the target image. The electron beam parameters were 5 kV at 1.6 nA beam current. Si (100) with 300 nm thermally grown SiO$_2$ was used as substrate material.

CONTENTS

Page

Functional Nanoscience
May 17th – 21st, 2010, Bozen, Italy

Functional Supramolecular Assemblies: First Glimpses and Upcoming Challenges

Yue Bing Zheng, Bala Krishna Pathem, J. Nathan Hohman and Paul S. Weiss*

California NanoSystems Institute and Departments of Chemistry & Biochemistry and Materials Science & Engineering, University of California, Los Angeles, CA 90095, U.S.A.

E-Mail: *psw@cnsi.ucla.edu

Received: 31st August 2010 / Published: 13th June 2011

Abstract

Functional supramolecular assemblies have emerged as candidates for nanosystems because of their potential for efficient, green, and cost-effective fabrication, as well as the capacity for tunability at molecular and atomic scales. Recent advances in the synthesis and characterization of supramolecular assemblies have already pointed to many opportunities ahead. While the practical integration of supramolecular assemblies into device manufacturing remains out of reach, learning to design the precise interactions between molecules and to control (and to predict) their structures will drive excitement, interest, and advances in bottom-up approaches to new materials and devices. This review discusses several key advances in the preparation of (and applications for) functional supramolecular assemblies, upcoming challenges, and our approaches toward the analysis of and precise control over functional supramolecular assemblies.

Introduction

A supramolecular assembly is an organized system of two or more molecules, where order originates from the weak, non-covalent binding interactions between the substituent parts [1, 2]. By tailoring molecular geometry and functionality, supramolecules may be designed and developed into functional nanosystems. While the lion's share of attention has been focused on the synthesis and characterization of various supramolecules, practical applications rely on the functional supramolecular assemblies that operate both independently and in concert [3]. It has been demonstrated that simple supramolecular components, after organization and assembly, can link molecular motions and reactions to complex macroscopic functions – including actuation and optical signal modulation [4 – 8]. Although the incorporation of functional supramolecular assemblies into modern fabrication methodologies remains a lofty goal, impressive progress has been made toward the development of functional nanomaterials and nanodevices [9]. Here, we highlight these materials and devices with focus on the upcoming challenges and our progress toward the precise control and analysis of functional supramolecular assemblies.

FUNCTIONAL SUPRAMOLECULAR ASSEMBLIES AT WORK

The understanding of how molecules spontaneously organize on surfaces has facilitated the design of functional supramolecular assemblies [10]. Exerting control over molecules at their fundamental scale has had a revolutionary impact on molecular electronics, nanoelectromechanical systems (NEMS), nanophotonics, and nanomedicine [4, 8, 11, 12]. A few examples of such designed assemblies are presented in the subsequent section to illustrate the power of these techniques.

Monolayers of bistable [2]rotaxanes for molecular plasmonics

Artificial molecular machines such as rotaxanes represent one class of supramolecules [1, 8, 13 – 15]. The rotaxanes may be described simply as a molecular dumbbell with mechanically interlocked rings (Fig. 1a) [4]. The bistable [2]rotaxanes employ the intramolecular recognition between the π-electron-poor tetracationic macrocycle cyclobis (paraquat-*p*-phenylene) ($CBPQT^{4+}$) rings and two π-electron-rich stations, namely 1,5-dioxynaphthalene (DNP) and the redox-active tetrathiafulvalene (TTF). The ring may be triggered to move from the TTF station to the DNP by oxidizing the TTF moiety, and subsequent TTF reduction reverses the motion. Before oxidation, the encircling ring is docked at the neutral TTF unit. Oxidation of the TTF unit to either the radical $TTF^{\bullet+}$ or dicationic TTF^{2+} lowers its affinity to the $CBPQT^{4+}$ ring, triggering the transit of the ring to the DNP station. Subsequent reduction of the $TTF^{\bullet+}$ or TTF^{2+} allows the ring to return to the neutral TTF station. This molecular motion is reversible, and may be repeated by cycling the TTF oxidation state.

Huang, Jensen, Stoddart, Weiss, and co-workers took advantage of the reversible, controllable mechanical motions in bistable [2]rotaxanes, and demonstrated a molecular-level plasmonic switch in which the localized surface plasmon resonances (LSPRs) of gold nanodisks could be reversibly controlled by mechanical motions in rotaxane molecules assembled on the disk surfaces [4]. Surface-plasmon-based nanophotonics, or "plasmonics", enables the development of photonic integrated circuits because of the capability for guiding and manipulating light at a scale smaller than its wavelength [16, 17]. Although major breakthroughs for passive components have been reported, including plasmonic waveguides, couplers, and lenses [18, 19], there have been few efforts to develop active components including switches and modulators [20 – 23].

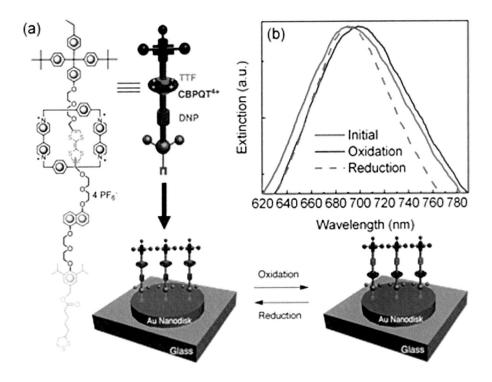

Figure 1. Plasmonic switches with rotaxanes as functional elements. **(a)** Functionalizing the rotaxane with the disulfide functional group allows it to be immobilized on the Au nanodisk surface. **(b)** Extinction spectra recorded from the rotaxane-derivatized gold nanodisks during the redox process provide a spectroscopic handle for monitoring the shuttling of the ring system. Reproduced with permission from reference [4]. Copyright 2009 American Chemical Society.

In their prototypical molecular plasmonic switches, Huang and co-workers have used the bistable [2]rotaxanes as the active component. Assembly on gold nanodisks via disulfide attachment to only one end of the rotaxane gives a non-centrosymmetric, upright orientation for the rotaxane molecules, and provides a strong chemical bond to the surface. A simple spectroscopic signature was used to track the ring position in this configuration. There is a strong visible absorption band centred at 820 nm while the ring is at the TTF station, a result of charge-transfer (CT) excitation between the highest occupied molecular orbital (HOMO) at the TTF and the lowest unoccupied molecular orbital (LUMO) of the CBPQT^{4+} ring. As before, oxidation causes the ring to shuttle from the TTF to the DNP station, which results in the disappearance of the absorption band centred at 820 nm. This change in the absorption band is associated with the change in the wavelength-dependent polarizability of rotaxane molecules, and thus the LSPRs of gold nanodisks. The extinction spectra for the rotaxane-derivatized gold nanodisks show that the oxidation causes the LSPRs peak to shift from 690.4 ± 0.2 to 699.9 ± 0.2 nm, while reduction caused the peak to blue shift back to its initial position (Fig. 1b). The reversible peak shift is reproducible for many redox cycles.

Monolayers of bistable [3]rotaxanes for nanoelectromechanical systems

As actuation materials, the rotaxanes have several advantages over biological molecular motors: high force density, wide range of pH stability (active at pH between 4 and 10), insensitivity to ambient temperature (operating between -30 to 100 °C), and may be triggered by multiple external stimuli [5]. These molecules have been engineered to generate precise, cooperative mechanical motion at the molecular scale, and represent the first step toward the engineering of systems that operate with the same elegance, efficiency, and complexity as the biological motors that drive the functions of the human body [24].

Huang, Stoddart, Weiss, and co-workers have demonstrated a mechanical actuator based on [3]rotaxane assemblies [6, 8]. Here, doubly bistable palindromic [3]rotaxane molecules self-assembled on a gold substrate have been used as an artificial molecular muscle (Fig. 2a). Before oxidation, the two TTF stations are nominally 8 nm apart. Oxidation causes the rings to move toward the centre of the molecule, with a spacing of approximately 4 nm. By attaching thiol functionality to each moving ring, rotaxanes could then be attached to cantilevers coated on only one side with gold (Fig. 2b). Oxidation triggers billions of these molecules simultaneously; the nanoscale motions collectively apply stress to one side of the cantilever. This stress is measured by the macroscopic deflection of the cantilevers. Figure 2c shows the time-dependent deflections of the rotaxane-modified microcantilever. This electrochemical method shows that the simultaneous 4-nm motions of individual molecules may be transduced into 300-nm deflections of a microcantilever (500 μm × 100 μm × 1 μm), illustrating the power of concerted nanoscale motions (Fig. 2b). The experimental observations, combined with elementary beam theory and analysis, indicate that the cumulative molecular-level movements within rotaxane assemblies bound on surfaces can be harnessed to perform macroscale mechanical work.

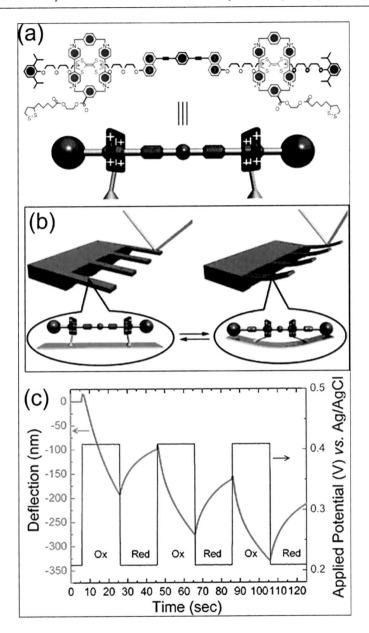

Figure 2. **(a)** Chemical structure and graphical representation of a disulfide-tethered bistable [3]rotaxane; **(b)** Schematic diagram of the deflections of cantilever by the cumulative molecular-level movements in [3]rotaxane assemblies bounded on the surface. Reproduced with permission from reference [6]. Copyright 2004 American Institute of Physics; **(c)** The deflection versus time (green line) of a cantilever coated on one side with the [3]rotaxanes on gold when subjected to a series of oxidation and reduction processes (purple line). Reproduced with permission from reference [8]. Copyright 2009 American Chemical Society.

Carbon nanotube yarns as multifunctional materials

Carbon nanotubes have attracted considerable interest due to their high tensile strength, as well as high electrical and thermal conductivities [25]. While individual carbon nanotubes have been successfully incorporated into electronic transistors, NEMS devices, and sensors [26–28], micro- and macroscale applications such as artificial muscles, conducting textiles, and fibre battery assemblies require trillions of long, oriented nanotubes [25, 29, 30]. Breakthroughs in the preparation of nanotube-based ropes have produced high-quality assemblies of both single-walled nanotubes (SWNTs) and multiwalled nanotubes (MWNTs) [29]. Baughman and co-workers spun and twisted forests of MWNTs into multi-ply, torque-stabilized yarns (Fig. 3) [25]. Due to the high surface-to-volume ratio of the MWNT and the high interfibre contact area per yarn volume, these yarns exhibit high twist retention and can be further knitted and knotted for a wider range of applications. The excellent material properties of individual nanotubes lead to yarns of superior mechanical properties with high strength, toughness, thermal stability, and abrasion resistance. The intertube mechanical coupling introduced by twisting also allows yarns to retain electronic connectivity between all component nanotubes, even during infiltration of poly(vinyl alcohol) (PVA).

With these enhanced mechanical and electrical properties, carbon nanotube yarns can serve as multifunctional materials and find important applications across many fields. Nanotube yarns could replace metal wires in electronic textiles with new functionalities, such as the ability to actuate as an artificial muscle and to store energy as a fibre supercapacitor or battery [25]. Furthermore, their conductivity can be exploited in soft, radio- and microwave-absorbing fabrics, personal protective equipment to protect against electrostatic discharge, and heat-resistant textiles [25].

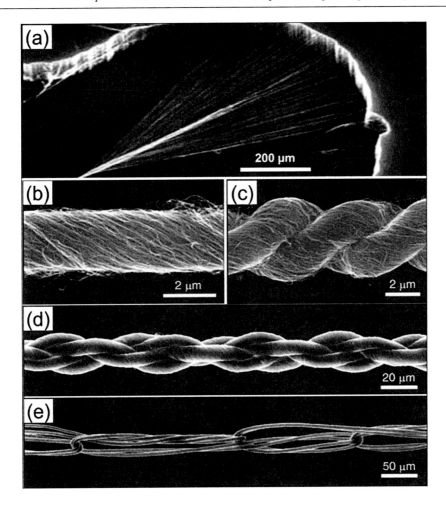

Figure 3. (a) A scanning electron microscopy (SEM) image showing a carbon nanotube yarn fabricated by simultaneous drawing and twisting from a vertically aligned multiwall nanotube (MWNT) forest; SEM images of **(b)** single, **(c)** two-ply, and **(d)** four-ply MWNT, as well as **(e)** knitted yarns. Reproduced with permission from reference [25]. Copyright 2004 American Association for the Advancement of Science.

UPCOMING CHALLENGES AND OPPORTUNITIES

Functional, oriented nanotube materials and robust supramolecular motors demonstrate the collective, macroscale effects of individual molecules acting in concert. [4, 8, 25, 29]. Despite research successes and growing momentum of the field over the last decade, there remain great opportunities for further advances. Controlling and measuring the physical, chemical, and electronic couplings between assembled molecules and between the assemblies and their substrates is a rich area for developments leading to a predictive understanding and the underlying design principles of optimized functional assemblies [11, 31 – 41]. Could arbitrary deflections of the cantilevers be achieved by controlling molecular orientations and interactions in the rotaxane assemblies? Can the response speed of the plasmonic switches be enhanced by engineering the rotaxane assemblies on the Au nanodisk surfaces? Is there a systematic approach toward optimization of the twist of MWCN yarns to have both maximized strength and conductivity? The study of single and precise supramolecular assemblies in well-defined environments offers insight into both couplings and functions. By isolating molecules and assemblies in a two-dimensional matrix, we can probe individual molecules, determine the role of the matrix, and design the matrix as an interacting part of the assembly [42 – 45]. In the following section, we report our progress in these directions, highlighting the capabilities of precise control and measurements we have and expect to have.

Using defects in self-assembled monolayers for precisely controlled molecular assemblies

Self-assembled monolayers (SAMs) are ultrathin films that spontaneously self-organize into one-molecule-thick structures on surfaces (Fig. 4a). The model system for SAMs has long been the *n*-alkanethiolate monolayers on gold; strong thiol-gold bonds and mobile complexes allow the alkyl chains to pack and to organize themselves rapidly into ordered structures [46 – 49]. Figure 4b shows a molecularly resolved scanning tunnelling microscopy (STM) image of a 1-dodecanethiolate SAM on Au{111}. A number of different types of defects exist in SAMs, and may be controlled by processing the film and substrate [50 – 52]. Defects are separated into two classes, substrate defects and monolayer defects, and are highlighted as indicated by arrows in Figure 4b. Gold substrate vacancy islands and gold terrace step edges are among the most prominent defects in the image. Step edges are a single-atom-high transition from one atomically flat terrace to another. Vacancy islands result from the removal of gold atoms during the self-assembly reaction, and manifest as one-atom-deep depressions in an otherwise atomically flat terrace. These two substrate defects can be treated identically, but are distinct features. The mismatch in substrate height prevents alkyl chains from maximizing tilt and thus van-der-Waals interactions. This loss of registry at the step edges results in voids that become active sites for molecular exchange and for the insertion of multiple (or larger) molecules. The mobility of gold atoms allows substrate vacancy islands to drift collectively relative to step edges; they will eventually merge with a terrace, a process that is greatly accelerated by elevating the deposition temperature and increasing deposition time.

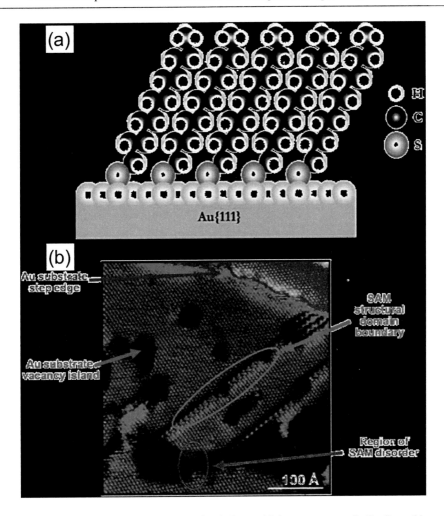

Figure 4. (a) Schematic of a SAM of 1-dodecanethiolate on an atomically flat gold substrate. Reproduced with permission from reference [51]. Copyright 2004 Elsevier Ltd; **(b)** A molecularly resolved STM image of a decanethiol SAM on Au{111}, indicating different types of defects. Reproduced with permission from reference [3]. Copyright 2008 American Chemical Society.

Monolayer defects include low-density, disordered regions and SAM structural domain boundaries. There are two types of domain boundaries: tilt domain boundaries and translational domain boundaries. Tilt domains result from a loss of registry due to different collective molecular tilts in ordered domains, while translational boundaries arise from mismatched positions of Au-S bonds between two domains. Structural domain boundaries are available sites for molecular exchange or for the insertion of *single* molecules or assemblies. Molecules can be inserted from the solution or vapour phases, or by contact via techniques such as micro-contact insertion printing [40, 53]. Isolating single molecules in well-defined environments is a powerful technique to control the orientation, isolation, and reactivity of individual molecules, and provides the basis for efficient bioselective capture surfaces (Fig. 5) [12, 54].

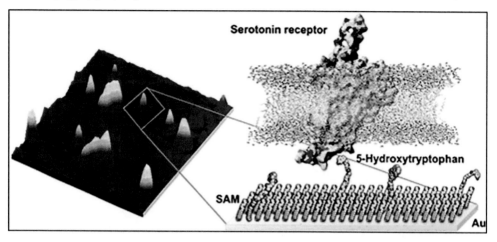

Figure 5. Capturing receptor proteins via surface-tethered small probes. A substrate is functionally modified with serotonin attached to oligo(ethylene glycol)-terminated alkanethiols self-assembled on Au{111}, and anti-5-hydroxytryptophan antibodies recognize the substrate, demonstrating bioavailability. Reproduced with permission from reference [12]. Copyright 2010 American Chemical Society.

Controlling azobenzene assemblies at surfaces

Photoisomerization about an azo functionality represents a fundamental molecular motor [21, 45, 55]. The coupling of external photoexcitation for the efficient control of surface-bound molecular assemblies remains a challenge [45]. This coupling is hampered by both the *quenching* of the excited states of the molecules by energy transfer to adjacent molecules and to the substrate, and to the inherent difficulty in measuring fast, transient events at a single-molecule scale. The molecule and its environment must be carefully designed to allow the tracking of single-molecule photoisomerization. First, molecules are isolated by insertion into the defects of SAMs, limiting their geometric degrees of freedom. The local environment has been prepared such that molecules tend to remain isolated, preventing energy transfer to neighbouring azobenzene units. The azobenzene moiety was attached to an ethoxy-butane-1-thiol tether. This serves to sever the functional part of the molecules from the substrate (**1** in Fig. 6). By controlling both the structure of the molecule and its local environment, we have been able to observe and to control the *trans-cis* reversible isomerization of individual azobenzene units using STM [45].

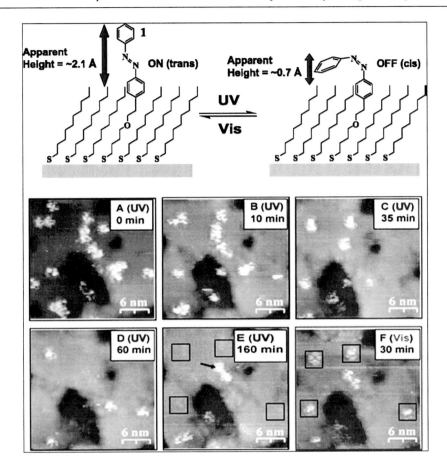

Figure 6. Isolated azobenzene molecules are embedded within 1-decanethiolate monolayer domains so as to eliminate motion about the S-Au bond. Reversible *trans-cis* photo-isomerization of the molecules by irradiating the surface with ultraviolet (~365 nm) and visible (~450 nm) light can be observed in sequence of images A-F. The isomerization is observed as an apparent height change in STM images. Imaging conditions: $V_{sample} = 1$ V; $I_{tunnel} = 2$ pA. Reproduced with permission from reference [45]. Copyright 2008 American Chemical Society.

As shown in Figure 6, without illumination, the azobenzenes were initially found to be in their thermodynamically stable *trans* state and appeared as 2.1 ± 0.3 Å protrusions over the 1-decanethiolate matrix in topographic STM images under the conditions shown. The surface was then exposed to UV light (~365 nm) and was simultaneously imaged by STM. Azobenzene molecules were found to isomerize to their *cis* state, characterized by an apparent height reduction of ~1.4 Å (Fig. 6B – 6E). After 160 min of UV illumination at ~12 mW/cm^2, greater than 90% of the azobenzene molecules had isomerized from *trans* to *cis*. Subsequent illumination with visible light (~450 nm) for 30 min at ~6 mW/cm^2 switched nearly 50% of the molecules back to their initial *trans* conformation (Figure 6F) [45].

When the azobenzenes were assembled into a chain, rather than isolated individually, in domain boundaries, a dramatic drop in photoisomerization efficiency was observed. Large steric hindrance and intermolecular electronic coupling could both be contributing to this effect. By increasing the chain length of the host matrix around the tethered azobenzenes we have shown that steric hindrance reduces the efficiency of photoisomerization. In order to determine the role of intermolecular excitation coupling, it is important to vary the intermolecular spacing between the active molecules precisely. This can be done through hierarchical assembly of specially designed molecules in order to elucidate the effects: the functionalization of active parts of the molecules when coupled to substrates, the intermolecular interactions in such assemblies, and the external environment. This will ultimately guide optimization of the assemblies in order to increase the switching efficiency and to have the molecules operate in concert for practical purposes.

Tracking individual rotaxane molecules on surfaces

Precise and simultaneous measurements of structure, dynamics, and indicators related to function are important for understanding molecular couplings in assemblies, and are powerful advantages of the STM [56, 57]. Weiss and co-workers and others have developed ultrastable and highly sensitive STMs that are able to image continuously a single region with minimized perturbations for several days [11, 58 – 63]. With advanced digital image processing techniques, data on single assemblies are automatically extracted from larger fields of view (b – e in Fig. 7), and are correlated with the local environment. Apparent height information is recorded relative to a local reference (Fig. 6) or adsorbate positions are determined to assign changes between images in order to follow surface dynamics (Fig. 7).

In the example of surface-bound rotaxanes, Weiss, Stoddart, and co-workers have been able to follow the motion of single rotaxane molecules with a stabilized STM under electrochemical control [64]. These observations were achieved by employing molecular designs that significantly reduced the mobility and enhanced the assembly of molecules in orientations conducive to direct measurement using STM. The results reveal molecular-level details of the station changes of surface-bound bistable rotaxane molecules, correlated with their different redox states (Fig. 7). The trajectories of the rings reveal that, once a bistable rotaxane molecule is adsorbed on a surface, the motion of the $CBPQT^{4+}$ ring relative to the dumbbell is affected by its local environment and the flexibility of the molecule. This study suggests that optimizing the design of rotaxane molecules with *rigid* dumbbells could further enhance the performance of rotaxane-based actuators (Fig. 2) [6, 8, 64].

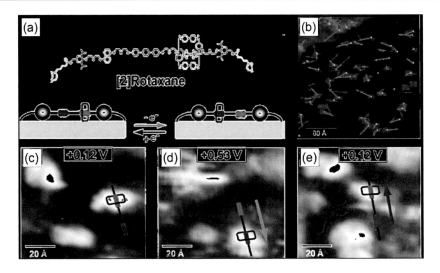

Figure 7. (a) Structure and motion of a bistable rotaxane molecule on Au{111}; **(b)** A STM image of rotaxanes adsorbed on Au{111} under 0.1 M HClO$_4$ solution. The three images with potentials +0.12, +0.53, and +0.12 V are superimposed and the protrusions (assigned as CBPQT^{4+} rings) have been marked. The trajectories of a large number of motions of the rings after potential steps from +0.12 to +0.53 V and back to +0.12 V are marked with blue and red lines, respectively; high-resolution image of an individual molecule **(c)** in its reduced state at +0.12 V with the ring at the TTF station; **(d)** upon oxidation (+0.12 to +0.53 V) the ring shuttling to the DNP station (blue arrow); **(e)** upon reduction (+0.53 to +0.12 V) with the ring returning to the TTF station (red arrow). Image conditions: $V_{bias} = 0.3$ V; $I_{tunnelling} = 2$ pA, Reproduced with permission from reference [64]. Copyright 2010 American Chemical Society.

CONCLUSIONS AND PROSPECTS

Although still in its nascent stage, research in the area of functional supramolecular assemblies has gained momentum over the past decade. A variety of supramolecules with precisely tailored structures and functions have been synthesized and characterized. Prototypical applications based on the collective effects in the functional supramolecular assemblies have emerged. However, many challenges remain in understanding and controlling a variety of molecular interactions in the assemblies, and in realizing coordinated action of functional molecules within such assemblies for practical devices. One effective way to address these challenges is to control assembly precisely and to measure the functional assemblies in well-controlled matrices. To this end, we have successfully developed procedures and tools for probing photoisomerization of azobenzene molecules and assemblies, and for tracking the motions of single rotaxane molecules. With new insights into building the assemblies gained through these studies, we will be able to design and to optimize the assemblies for greater function at the nanoscale.

ACKNOWLEDGMENTS

We thank the NSF-supported Center for Nanoscale Science (MRSEC), the Department of Energy (Grant No. DE-FG02 – 07ER15877), and the Kavli Foundation for support of the work described here. We acknowledge Profs. David Allara, Tony Jun Huang, Lasse Jensen, Fraser Stoddart, and James Tour for their collaborations in these efforts.

REFERENCES

[1] Li, D.B., Paxton, W.F., Baughman, R.H., Huang, T.J., Stoddart, J.F., Weiss, P.S. (2009) *MRS Bull.* **34**:671.
doi: 10.1557/mrs2009.179.

[2] Barth, J.V., Weckesser, J., Cai, C.Z., Gunter, P., Burgi, L., Jeandupeux, O., Kern, K. (2000) *Angew. Chem. Int. Edit.* **39**:1230.
doi: 10.1002/(SICI)1521-3773(20000403)39:7<1230::AID-ANIE1230>3.0.CO;2-I.

[3] Weiss, P.S. (2008) *Acc. Chem. Res.* **41**:1772.
doi: 10.1021/ar8001443.

[4] Zheng, Y.B., Yang, Y.W., Jensen, L., Fang, L., Juluri, B.K., Flood, A.H., Weiss, P.S., Stoddart, J.F., Huang, T.J. (2009) *Nano Lett.* **9**:819.
doi: 10.1021/nl803539g.

[5] Huang, T.J., Flood, A.H., Brough, B., Liu, Y., Bonvallet, P.A., Kang, S.S., Chu, C.W., Guo, T.F., Lu, W.X., Yang, Y., Stoddart, J.F., Ho, C.M. (2006) *IEEE Trans. Autom. Sci. Eng.* **3**:254.
doi: 10.1109/TASE.2006.875543.

[6] Huang, T.J., Brough, B., Ho, C.M., Liu, Y., Flood, A.H., Bonvallet, P.A., Tseng, H.R., Stoddart, J.F., Baller, M., Magonov, S. (2004) *Appl. Phys. Lett.* **85**:5391.
doi: 10.1063/1.1826222.

[7] Huang, T.J. (2008) *MRS Bull.* **33**:226.
doi: 10.1557/mrs2008.48.

[8] Juluri, B.K., Kumar, A.S., Liu, Y., Ye, T., Yang, Y.W., Flood, A.H., Fang, L., Stoddart, J.F., Weiss, P.S., Huang, T.J. (2009) *ACS Nano* **3**:291.
doi: 10.1021/nn8002373.

[9] Michl, J., Sykes, E.C.H. (2009) *ACS Nano* **3**:1042.
doi: 10.1021/nn900411n.

[10] Barth, J.V., Costantini, G., Kern, K. (2005) *Nature* **437**:671.
doi: 10.1038/nature04166.

[11]   Donhauser, Z.J., Mantooth, B.A., Kelly, K.F., Bumm, L.A., Monnell, J.D., Stapleton,
        J.J., Price, D.W., Rawlett, A.M., Allara, D.L., Tour, J.M., Weiss, P.S. (2001) *Science*
        292:2303.
        doi: 10.1126/science.1060294.

[12]   Vaish, A., Shuster, M.J., Cheunkar, S., Singh, Y.S., Weiss, P.S., Andrews, A.M.
        (2010) *ACS Chem. Neurosci.* **1**:495.
        doi: 10.1021/cn1000205.

[13]   Stoddart, J.F., Tseng, H.R. (2002) *Proc. Natl. Acad. Sci. U.S.A.* **99**:4797.
        doi: 10.1073/pnas.052708999.

[14]   Green, J.E., Choi, J.W., Boukai, A., Bunimovich, Y., Johnston-Halperin, E., DeIon-
        no, E., Luo, Y., Sheriff, B.A., Xu, K., Shin, Y.S., Tseng, H.R., Stoddart, J.F., Heath,
        J.R. (2007) *Nature* **445**:414.
        doi: 10.1038/nature05462.

[15]   Flood, A.H., Ramirez, R.J.A., Deng, W.Q., Muller, R.P., Goddard, W.A., Stoddart,
        J.F. (2004) *Aust. J. Chem.* **57**:301.
        doi: 10.1071/CH03307.

[16]   Engheta, N. (2007) *Science* **317**:1698.
        doi: 10.1126/science.1133268.

[17]   Atwater, H.A., Maier, S., Polman, A., Dionne, J.A., Sweatlock, L. (2005) *MRS Bull.*
        30:385.

[18]   Fang, N., Lee, H., Sun, C., Zhang, X. (2005) *Science* **308**:534.
        doi: 10.1126/science.1108759.

[19]   Bozhevolnyi, S.I., Volkov, V.S., Devaux, E., Laluet, J.Y., Ebbesen, T.W. (2006)
        Nature **440**:508.
        doi: 10.1038/nature04594.

[20]   Krasavin, A.V., Zheludev, N.I. (2004) *Appl. Phys. Lett.* **84**:1416.
        doi: 10.1063/1.1650904.

[21]   Hsiao, V.K.S., Zheng, Y.B., Juluri, B.K., Huang, T.J. (2008) *Adv. Mater.* **20**:3528.
        doi: 10.1002/adma.200800045.

[22]   Chang, D.E., Sorensen, A.S., Demler, E.A., Lukin, M.D. (2007) *Nat. Phys.* **3**:807.
        doi: 10.1038/nphys708.

[23]   Andrew, P., Barnes, W.L. (2004) *Science* **306**:1002.
        doi: 10.1126/science.1102992.

[24]   Kottas, G.S., Clarke, L.I., Horinek, D., Michl, J. (2005) *Chem. Rev.* **105**:1281.
        doi: 10.1021/cr0300993.

[25]   Zhang, M., Atkinson, K.R., Baughman, R.H. (2004) *Science* **306**:1358.
        doi: 10.1126/science.1104276.

[26]   Tans, S.J., Verschueren, A.R.M., Dekker, C. (1998) *Nature* **393**:49.
        doi: 10.1038/29954.

[27]   Spinks, G.M., Wallace, G.G., Fifield, L.S., Dalton, L.R., Mazzoldi, A., De Rossi, D.,
        Khayrullin, II, Baughman, R H. (2002) *Adv. Mater.* **14**:1728.

[28]   Stampfer, C., Helbling, T., Obergfell, D., Schoberle, B., Tripp, M.K., Jungen, A.,
        Roth, S., Bright, V.M., Hierold, C. (2006) *Nano Lett.* **6**:233.
        doi: 10.1021/nl052171d.

[29]   Zhang, M., Fang, S.L., Zakhidov, A.A., Lee, S.B., Aliev, A.E., Williams, C.D.,
        Atkinson, K.R., Baughman, R.H. (2005) *Science* **309**:1215.
        doi: 10.1126/science.1115311.

[30]   Aliev, A.E., Oh, J.Y., Kozlov, M.E., Kuznetsov, A.A., Fang, S.L., Fonseca, A.F.,
        Ovalle, R., Lima, M.D., Haque, M. H., Gartstein, Y.N., Zhang, M., Zakhidov, A.A.,
        Baughman, R.H. (2009) *Science* **323**:1575.
        doi: 10.1126/science.1168312.

[31]   Lewis, P.A., Inman, C.E., Maya, F., Tour, J.M., Hutchison, J.E., Weiss, P.S. (2005) *J.
        Am. Chem. Soc.* **127**:17421.
        doi: 10.1021/ja055787d.

[32]   Donhauser, Z.J., Price, D.W., Tour, J.M., Weiss, P.S. (2003) *J. Am. Chem. Soc.*
        125:11462.
        doi: 10.1021/ja035036g.

[33]   Lewis, P.A., Inman, C.E., Yao, Y.X., Tour, J.M., Hutchison, J.E., Weiss, P.S. (2004)
        J. Am. Chem. Soc. **126**:12214.
        doi: 10.1021/ja038622i.

[34]   Moore, A.M., Dameron, A.A., Mantooth, B.A., Smith, R.K., Fuchs, D.J., Ciszek,
        J.W., Maya, F., Yao, Y.X., Tour, J.M., Weiss, P.S. (2006) *J. Am. Chem. Soc.*
        128:1959.
        doi: 10.1021/ja055761m.

[35]   Moore, A.M., Mantooth, B.A., Dameron, A.A., Donhauser, Z.J., Lewis, P.A., Smith,
        R.K., Fuchs, D.J., Weiss, P.S. (2008) *Front. Mat. Res.* **10**:29.
        doi: 10.1007/978-3-540-77968-1_3.

[36]   Dameron, A.A., Ciszek, J.W., Tour, J.M., Weiss, P.S. (2004) *J. Phys. Chem. B* **108**:16761.
       doi: 10.1021/jp049442d.

[37]   Moore, A.M., Mantooth, B.A., Donhauser, Z.J., Maya, F., Price, D.W., Yao, Y.X., Tour, J.M., Weiss, P.S. (2005) *Nano Lett.* **5**:2292.
       doi: 10.1021/nl051717t.

[38]   Mantooth, B.A., Weiss, P.S. (2003) *Proc. IEEE* **91**:1785.
       doi: 10.1109/JPROC.2003.818320.

[39]   Love, J.C., Estroff, L.A., Kriebel, J.K., Nuzzo, R.G., Whitesides, G.M. (2005) *Chem. Rev.* **105**:1103.
       doi: 10.1021/cr0300789.

[40]   Mullen, T.J., Srinivasan, C., Shuster, M.J., Horn, M.W., Andrews, A.M., Weiss, P.S. (2008) *J. Nanopart. Res.* **10**:1231.
       doi: 10.1007/s11051-008-9395-y.

[41]   Mullen, T.J., Dameron, A.A., Andrews, A.M., Weiss, P.S. (2007) *Aldrichim. Acta* **40**:21.

[42]   Bumm, L.A., Arnold, J.J., Cygan, M.T., Dunbar, T.D., Burgin, T.P., Jones, L., Allara, D.L., Tour, J.M., Weiss, P.S. (1996) *Science* **271**:1705.
       doi: 10.1126/science.271.5256.1705.

[43]   Cygan, M.T., Dunbar, T.D., Arnold, J.J., Bumm, L.A., Shedlock, N.F., Burgin, T.P., Jones, L., Allara, D.L., Tour, J.M., Weiss, P.S. (1998) *J. Am. Chem. Soc.* **120**:2721.
       doi: 10.1021/ja973448 h.

[44]   Weck, M., Jackiw, J.J., Rossi, R.R., Weiss, P.S., Grubbs, R.H. (1999) *J. Am. Chem. Soc.* **121**:4088.
       doi: 10.1021/ja983297y.

[45]   Kumar, A.S., Ye, T., Takami, T., Yu, B.C., Flatt, A.K., Tour, J.M., Weiss, P.S. (2008) *Nano Lett.* **8**:1644.
       doi: 10.1021/nl080323+.

[46]   Nuzzo, R.G., Allara, D.L. (1983) *J. Am. Chem. Soc.* **105**:4481.
       doi: 10.1021/ja00351a063.

[47]   Stapleton, J.J., Harder, P., Daniel, T.A., Reinard, M.D., Yao, Y.X., Price, D.W., Tour, J.M., Allara, D.L. (2003) *Langmuir* **19**:8245.
       doi: 10.1021/la035172z.

[48]   Tamchang, S.W., Biebuyck, H.A., Whitesides, G.M., Jeon, N., Nuzzo, R.G. (1995) *Langmuir* **11**:4371.
        doi: 10.1021/la00011a033.

[49]   Vericat, C., Andreasen, G., Vela, M.E., Martin, H., Salvarezza, R.C. (2001) *J. Chem. Phys.* **115**:6672.
        doi: 10.1063/1.1403000.

[50]   Hohman, J.N., Claridge, S.A., Kim, M., Weiss, P.S. (2010) *Mat. Sci. Eng. R*
        doi: 10.1016/j.mser.2010.06.008.

[51]   Smith, R.K., Lewis, P.A., Weiss, P.S. (2004) *Prog. Surf. Sci.* **75**:1.
        doi: 10.1016/j.progsurf.2003.12.001.

[52]   Saavedra, H.M., Mullen, T.J., Zhang, P.P., Dewey, D.C., Claridge, S.A., Weiss, P.S. (2010) *Rep. Prog. Phys.* **73**:036501.
        doi: 10.1088/0034-4885/73/3/036501.

[53]   Mullen, T.J., Srinivasan, C., Hohman, J.N., Gillmor, S.D., Shuster, M.J., Horn, M.W., Andrews, A.M., Weiss, P.S. (2007) *Appl. Phys. Lett.* **90**.
        doi: 10.1021/jp063309z.

[54]   Shuster, M.J., Vaish, A., Szapacs, M.E., Anderson, M.E., Weiss, P.S., Andrews, A.M. (2008) *Adv. Mater.* **20**:164.
        doi: 10.1002/adma.200700082.

[55]   Katsonis, N., Lubomska, M., Pollard, M.M., Feringa, B.L., Rudolf, P. (2007) *Prog. Surf. Sci.* **82**:407.
        doi: 10.1016/j.progsurf.2007.03.011.

[56]   McCarty, G.S., Weiss, P.S. (1999) *Chem. Rev.* **99**:1983.
        doi: 10.1021/cr970110x.

[57]   Moore, A.M., Weiss, P.S. (2008) *Annu. Rev. Anal. Chem.* **1**:857.
        doi: 10.1146/annurev.anchem.1.031207.112932.

[58]   Stranick, S.J., Kamna, M.M., Weiss, P.S. (1994) *Rev. Sci. Instrum.* **65**:3211.
        doi: 10.1063/1.1144551.

[59]   Ferris, J.H., Kushmerick, J.G., Johnson, J.A., Youngquist, M.G.Y., Kessinger, R.B., Kingsbury, H.F., Weiss, P.S. (1998) *Rev. Sci. Instrum.* **69**:2691.
        doi: 10.1063/1.1149000.

[60]   Bumm, L.A., Arnold, J.J., Dunbar, T.D., Allara, D.L., Weiss, P.S. (1999) *J. Phys. Chem. B* **103**:8122.
        doi: 10.1021/jp9921699.

[61] Besenbacher, F., Lauritsen, J.V., Wendt, S. (2007) *Nano Today* **2**:30.
doi: 10.1016/S1748-0132(07)70115-9.

[62] Otero, R., Hummelink, F., Sato, F., Legoas, S.B., Thostrup, P., Laegsgaard, E., Stensgaard, I., Galvao, D.S., Besenbacher, F. (2004) *Nat. Mater.* **3**:779.
doi: 10.1038/nmat1243.

[63] Linderoth, T.R., Horch, S., Laegsgaard, E., Stensgaard, I., Besenbacher, F. (1998) *Surf. Sci.* **404**:308.
doi: 10.1016/S0039-6028(97)01054-6.

[64] Ye, T., Kumar, A.S., Saha, S., Takami, T., Huang, T.J., Stoddart, J.F., Weiss, P.S. (2010) *ACS Nano* **4**:3697.
doi: 10.1021/nn100545r.

Catalytic Nanomotor Function and Locomotion Physics

Jonathan D. Posner[1,*], Jeffrey L. Moran[1], Joseph Wang[2] and Philip Wheat[1]

[1]Mechanical & Aerospace Engineering, Chemical Engineering, Arizona State University, Tempe, AZ 85287, U.S.A.

[2]Department of Nanoengineering, University of California San Diego, La Jolla, CA 92093, U.S.A.

E-Mail: *jposner@asu.edu

Received: 20th August 2010 / Published: 13th June 2011

Introduction

In nature, microorganisms propel themselves through fluid media by mechanical deformation of solid appendages using energy they harvest from their local environment [4, 24, 36, 40]. Alternatively, Mitchell proposed that an asymmetric ion flux on a bacterium's surface could generate electric fields that drive locomotion via self-electrophoresis [27, 28, 41]. This proposed mechanism grew out of research on endogenous bioelectric fields which provided motive force in morphogenesis [13, 17, 25, 31]. Lund and Jaffe realized that asymmetric currents on the surface of a cell or embryo could generate bioelectric fields that alter the shape of cells and tissues as well as drive the transport of biomolecules. Roughly thirty years after Mitchell's hypothesis, Waterbury et al. [39] discovered that cyanobacteria are able to swim at speeds up to 25 μm s^{-1} without flagella, suggesting that they move by a non-mechanical propulsion mechanism. This theory was explored by Anderson (and later Lammert et al. [22]) who hypothesized that Waterbury's flagella-less cyanobacteria may swim due to Mitchell's self-electrophoretic mechanism [1]. Mitchell's self-electrophoresis mechanism and the subsequent analysis by Anderson were rejected for cyanobacteria by Pitta and Berg [35].

Recent advances in nanofabrication have enabled the engineering of synthetic analogues to those proposed by Mitchell that swim due to asymmetric ion flux mechanisms. Several realizations of synthetic swimmers, or artificial nanomotors, have been demonstrated. A common theme among these artificial swimmers is the production of species concentration gradients by asymmetric chemical or electrochemical reactions occurring on the surface of Janus particles [8, 11, 12, 16]. Electrochemical reactions have also been shown to drive the motion of colloidal particles. The particles effectively act as short-circuited electrochemical cells. Mano and Heller [26] demonstrated the propulsion of a carbon fibre by redox reactions of glucose and oxygen occurring on opposite sides of the fibre. Propulsion due to electrochemical reactions has also been demonstrated with decomposition of hydrogen peroxide. The peroxide undergoes two electrochemical half-reactions before finally yielding oxygen and water. Electrochemical peroxide decomposition has been reported to generate fluid flows above concentric electrodes [19] and in between interdigitated microelectrodes [34] as well as to propel rotating micro-gears made from platinum and gold [7] and bimetallic nanorod motors [23, 33, 38].

Bimetallic nanorods are high aspect ratio metallic cylinders, typically $2\,\mu m$ in length and ≈ 300 nm in diameter. Swimming bimetallic rods were introduced to the community by Paxton *et al.* [32]. Laocharoensuk *et al.* [23] added hydrazine to the hydrogen peroxide solutions and incorporated carbon nanotubes into the platinum segments of Pt/Au rods to obtain nanomotor velocities in excess of $200\,\mu m\ s^{-1}$ (100 body lengths per second). Kagan *et al.* [18] demonstrated that the nanomotors' velocity strongly depends upon the presence of ions in the solution. They showed that for most electrolytes, nanomotor speed decreases with increasing salt concentration; however, silver ions caused a fivefold increase in the nanomotors' speed. A nanomotor's velocity can be externally modulated by heat pulses [2] and by electrochemically controlling the local concentration of hydrogen peroxide and dissolved molecular oxygen [6].

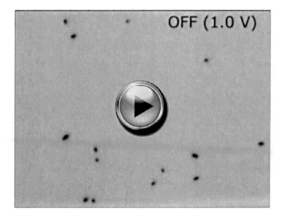

Movie 1. Electrochemical Switching of Nanomotors Motion.

Directional control of the nanomotors' motion has been achieved using external magnetic fields by incorporating a ferromagnetic segment, such as nickel, between the platinum and gold segments [5, 19]. We and others have shown that nanomotors can pick up, haul, and release micron-scale cargo [5, 37]. By hauling cargo of various sizes, we calculated that these nanomotors produce roughly 0.1 pN of force. In the case of bimetallic nanomotors, the mechanism by which asymmetric reactions generate forces and locomotion is not well understood. Several theories have been proposed. In their initial paper introducing the nanomotors [32], Paxton and co-workers proposed a propulsion mechanism based on the formation of gradients in interfacial tension. According to this mechanism, oxygen produced at the platinum end causes a local decrease in the interfacial tension between the aqueous solution and the gas-coated rod. The interfacial tension gradient then drives a Marangoni flow that results in a net propulsive force that drives particle motion. Although this theory predicts a propulsive force of the appropriate magnitude, it makes no predictions regarding the direction of particle motion. Paxton *et al.* [33] considered the possibility of motion due to a concentration gradient in molecular oxygen drives the particle due to diffusiophoresis. They found that the predicted diffusiophoretic velocity was much smaller than that observed in experiments and in the wrong direction. A later study showed that fluid could be pumped as a result of flows induced by electrochemical reactions occurring on concentric metallic electrodes [20]. This system was later studied analytically and experimentally using the Poisson, advection-diffusion, and Helmholtz-Smoluchowski equations [21].

The physics of this mechanism is similar to the self-electrophoretic mechanism proposed by Mitchell [27]. Although there is a growing consensus that Mitchell's self-electrophoretic mechanism is important in the motion of bimetallic nanomotors, many key physical details of the remain unexplained.

Here we are interested in locomotion of particles that are driven due to concentration and potential gradients that are generated by reactions occurring on the particle surface. We present some experimental results describing the controlled motion of catalytic nanomotors as well as some numerical simulations on the physical mechanisms underlying the locomotion of particles due to asymmetric reactions. Since we solve the full equations for conservation of mass, species, and momentum as well as Poisson's equation for electric potential, our model incorporates all effects due to diffusiophoresis, chemiphoresis, and electrophoresis. The simulations provide quantitative predictions of the nanomotor's swimming velocity for a given surface charge and reaction rate. We observe excellent agreement between experiments with our scaling analysis and numerical models.

DESCRIPTION OF THE PROBLEM

Figure 1 shows a schematic of a platinum-gold nanorod in an aqueous hydrogen peroxide solution. If the rod is immersed in a hydrogen peroxide solution, catalytic electrochemical reactions occur on its surface. On the platinum end, an oxidation reaction decomposes peroxide into protons, electrons, and oxygen molecules. The oxidation reaction results in a local depletion of the peroxide fuel and abundance of protons and oxygen. The electrons conduct through the rod from platinum to gold. At the gold end, two reduction reactions are possible which both result in the production of water. First, protons, electrons, and peroxide react to form water. Second, molecular oxygen, protons, and electrons can react to generate water. At the Au end there is a net depletion of protons and, to a lesser extent, peroxide and molecular oxygen. In this way, the nanomotor effectively functions as a short-circuited galvanic cell which generates a surface flux of protons at the anode (Pt end) and consumes protons at the cathode (Au end). This basic description of the reaction mechanism was used by Wang *et al.* [38] to successfully predict the direction of motion of a nanomotor composed of any two of six noble metals.

Here we provide a model that shows that the locomotion is driven by fluid slip around the nanomotor caused by self-generated electrical body forces in a mechanism called reaction induced charge auto-electrophoresis (RICA) [29, 30]. The respective excess and depletion of protons at the anode and cathode generate an electric field, E. Electric body forces, of the form $\rho_e E$, arise due to the coupling of free charge density, ρ_e, with tangential electric fields which are generated by the asymmetric charge density distribution induced by the asymmetric surface reactions. The charge density results from the asymmetry in cation concentration (due to the reactions, as shown in Figure 1) as well as from the diffuse portion of the electric double layer (EDL) that forms due to the rod's native surface charge (not shown in Figure 1). The electric body forces drive fluid motion and thus nanomotor motion. Ultimately, net locomotion is driven by a coupling of the diffuse charge in the EDL that shields native surface charge on nanomotor surface and electric fields that are driven by the asymmetric distribution of reaction products established by the heterogeneous surface reactions.

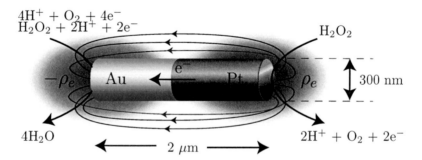

Figure 1. Schematic of Pt/Au catalytic nanomotor in hydrogen peroxide solution showing typical dimensions, electrochemical reactions, approximate charge density distribution, and approximate electric field lines. As shown here, the rod's motion is directed to the right.

EXPERIMENTS

Experimental Methodology

The magnetic nanomotors were prepared by electrodepositing the corresponding metals or hybrid metal-CNT composite into a porous alumina membrane template (Catalogue no. 6809 – 6022; Whatman, Maidstone, U.K.). A thin gold film was first sputtered on the branched side of the membrane to serve as a working electrode. The membrane was assembled in a plating cell with aluminium foil serving as an electrical contact for the subsequent electro-deposition. In order to synthesize well-shaped cylindrical nanomotors, a sacrificial silver layer was electrodeposited into the branched area ($1 - 2\,\mu m$ thickness) of the membrane using a silver plating solution (1025 RTU@4.5 Troy/Gallon; Technic Inc., Anaheim, CA, U.S.A.) and a total charge of 2 coulombs (C) at -0.9 V (vs. Ag/AgCl, in connection to a Pt wire counter electrode). This was followed by the electro-deposition of gold (0.75 C) from a gold plating solution (Orotemp 24 RTU RACK; Technic Inc.) at -0.9 V (vs. Ag/AgCl). Following the plating of the first gold segment (0.75 C), nickel was electro-deposited from a nickel plating solution 20 g l^{-1} $NiCl_2 \cdot 6\,H_2O$, 515 g l^{-1} $Ni(H_2NSO_3)_2 \cdot 4\,H_2O$, and 20 g l^{-1} H_3BO_3 (buffered to pH 3.4)] at -1.0 V (vs. Ag/AgCl). A total charge of 2.0 C was used for plating nickel. The second gold segment (0.75 C) was then deposited. Subsequently, platinum-CNT were deposited galvanostatically at -2 mA for 50 min from a platinum plating solution containing 0.50 mg/ml of CNT, along with 0.1 wt% Nafion and 2 mM 4-nitrobenzenediazonium tetrafluoroborate, respectively. After depositing the nanomotors, the membrane was removed from the plating cell and rinsed thoroughly with nanopure water to remove all residues. The sputtered gold layer and the silver layer were simultaneously removed by mechanical polishing using cotton tip applicators soaked with 35% HNO_3 for ca. 5 min to ensure complete silver dissolution. The nanomotors were then released by immersing the membrane in 3 M NaOH for 30 min. These nanowires were collected by centrifugation at 10,000 rpm for 5 min and washed repeatedly with nanopure water (18.2 M cm) until a neutral pH was achieved. Between washing steps the nanowire solution was mixed and briefly sonicated (several seconds) to ensure the complete dispersion of nanowires in the washing water and hence the removal of salt residuals entrapped in the nanowire aggregate after centrifugation.

Special attention was paid to the nanowires being washed directly before testing and suspended in freshly obtained nanopure water due to significant deceleration of the nanomotors speed in the presence of salt ions. All nanomotors were stored in nanopure water at room temperature and their speed tested within a day of synthesis.

The nanomotor suspensions were diluted in water and mixed with hydrogen peroxide solution to obtain roughly 5 wt% H_2O_2, unless noted. The real-time movement of the nanomotors was visualized using white light transmission microscopy at 200 × total magnification (Nikon Instrument Inc., Eclipse80i, Melville, NY, U.S.A.) equipped with a Photometrics CoolSnap CF camera (Roper Scientific, Duluth, GA, U.S.A.). The nanomotors where controlled using a 9.5 mm cube-shaped Neodymium (NdFeB) magnet placed at the centre of the top face of a larger 25.4 mm NdFeB cube shaped magnet (both 1.32 Tesla from K&J Magnetics Inc., Jamison, PA, U.S.A.). The magnet pair was fixed to a custom magnet holder, and attached directly to the microscope condenser stage thus aligning the centre of the magnets with the optical axis.

The magnetic poles (from north to south) of the large and small magnets were opposite in direction while both were perpendicular to the optical axis of the microscope. The super-position of the magnetic fields of the two magnets allowed for a weak parallel magnetic field at a distance greater than 2 mm and a strong perpendicular field ca. 1 mm. For cargo pickup and release experiments a diluted suspension of nanomotors and magnetic-nanoparticles coated polystyrene beads (Spherotech, Inc., Libertyville, IL, U.S.A.) was prepared. A 10 mm cube-shaped magnet was placed on the microscope stage next to a microcapillary tube, containing the nanomotors and spherical cargo. To measure the speeds of a nanomotor before picking up, during transporting and after dropping off the cargo, the nanomotor was magnetically directed toward an unbound magnetic microsphere, captured and transported it for ca. 5 seconds and finally releasing the cargo via a fast reversal of the magnetic poles.

Results

We demonstrate the directed motion of catalytic nanomotors and their cargo transport and manipulation capabilities along predetermined paths within micro-channel networks. We eliminate the requirement for functionalization of the nanowire by making use of the magnetic properties of a nickel segment for guiding and sorting of nanomotors through various junctions of the microfluidic network at speeds up to 40 μm s^{-1}. In addition, the magnetic segments enable dynamic loading, transport, and release of spherical magnetic cargo. The ability of a nanomotor to travel on a predetermined path is an important aspect of the integration of nanomotors into microfluidic networks. Directed motion of chemically-powered Au/Ni/Au/Pt-CNT nanowires within a micro-channel network is illustrated in Figure 2. A weak external magnet was used for sorting the nanomotors in each of the junctions of the microfluidic network, without contributing to the nanomotors speed. For example, Figure 2A illustrates a 2 μm long Au/Ni/Au/Pt-CNT nanomotor travelling in an hourglass pattern along a 700 μm long preselected path in a symmetric central portion of the micro-channel network over a 70 second period. Figure 2B illustrates the directional manip-ulation of the nanomotors within the micro-channel network. Four predetermined paths (a-d), starting from the same starting point (SP), illustrate the dexterity of the CNT-based nanomotors and the magnetic sorting in different junctions. It should be noted that the same nanowire was travelling continuously in all four paths of Figure 2B (as well as returning to the same starting point) for more than an hour.

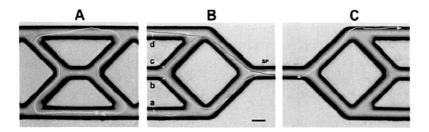

Figure 2. Tracks of directed nanomotor movement within a microfabricated channel network. **(A)** Motion of an Au/Ni/Au/Pt-CNT nanomotor maneuvering around the central portion of a PDMS microstructure. **(B)** Motion of the same nanowire over four different paths (a-d) beginning from the same starting point (SP). **(C)** Motion of a nanomotor in a mixed peroxide/hydrazine fuel. Fuels, 5 wt% H_2O_2 **(A, B)**; 2.5 wt% $H_2O_2/0.15$ wt% hydrazine **(C)**. Scale bar 25 μm. Reprinted with permission from [5]. Copyright 2010 American Chemical Society.

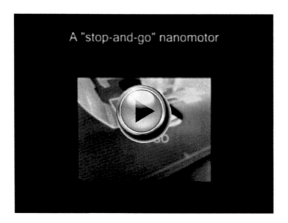

Movie 2. A "stop-and-go" nanomotor.

The magnetic properties of the Au/Ni/Au/Pt-CNT nanomotor can also be exploited for executing a controlled pick up and release of the magnetic-spherical cargo as shown in Figure 3 [5]. On-demand cargo pick up is accomplished by a weak magnetic interaction between the superparamagnetic spherical particle and the ferromagnetic nickel segment in the nanomotor. The nanomotor is directed close to the spherical particle. When the nano-motor is in close proximity to the particle, (ca. one particle diameter) the magnetic fields of the particle and nanomotor are sufficient to bring them together. The spherical particle-nanomotor complex continues to move, powered by RICA, along the path dictated by the external magnetic field. Release is accomplished by exploiting the weak interaction between the nickel segment within the nanomotor and the magnetic polystyrene beads. A fast change in the direction of the nanomotor (through rapid 180 degree rotation of the external magnet) releases the magnetic cargo when the viscous fluid drag on the particle exceeds the magnetic force between the nanowire and the particle. This cargo manipulation functionality could be used for controlled assembly, autonomous microdevices, or drug delivery, for example.

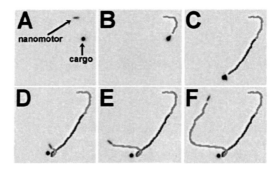

Figure 3. Sequential micrographs **(A-F)** of cargo pickup, transport, and release by an Au/Ni/Au/Pt-CNT nanomotor. Green and red traces denote the path traveled by the nanomotor and the cargo-loaded nanomotor, respectively: **(A and B)** the manipulation toward and pickup of cargo by a nanomotor; **(C)** subsequent transport of the cargo by the nanomotor; **(D)** release of the cargo; **(E and F)** motion of the nanomotor after the cargo release. Reprinted with permission from [5]. Copyright 2008 American Chemical Society.

Movie 3. In-dive cargo pick-up by magnetic controlled synthetic nanomotors.

We measure the velocity of the nanomotor-particle complexes and find that that they depend on the size of the cargo. Figure 4 shows the velocity of the complex as a function of the size of the particle diameter. This plot shows that the nanomotor velocity decreases from 11 to 4 μm s^{-1} upon increasing the cargo size from 1.3 to 4.3 μm, respectively.

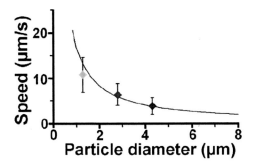

Figure 4. Nanomotor-sphere complex velocity as a function of the attached sphere diameter. The solid line is a simple theory based on the Stokes drag of the sphere. Reprinted with permission from [5]. Copyright 2010 American Chemical Society.

We expect that the complexes' velocity to depend on the size of particle because the speed of the terminal velocity of the nanomotor is reached when the propulsive force of the electro-kinetic locomotion is balanced with the hydrodynamic Stokes drag as,

$$F_d = F_{nm} = 6\pi\mu U a,$$

where μ is the fluid viscosity, U the velocity of the nanomotor complex, a is the particle radius, F_d is the drag force, and F_{nm} is the force the nanomotor generates due to RICA. Using this equation, we can make a theoretical prediction of the complexes' velocity as a function of the cargo diameter. This prediction is shown along with the experimental data in Figure 4. This theory requires a single fitting parameter, namely the force generated by the nanomotor. We find that fitting this theory to the data results in a nanomotor force of 0.16 pN. Smaller forces (0.04 pN) have been reported for conventional (undoped) Pt-Au nanowires. The larger cargo-towing forces of the CNT-based nanomotors allow transport of larger cargo at a faster speed compared to common Pt-Au nanowires (e. g., 4 μm/s for the 4.3 μm cargo compared to 3.5 μm/s for the 2.1 μm cargo, respectively).

We have also measured the velocity of nanomotors as a function of peroxide concentration. Figure 5 shows the velocity of undoped nanomotors as a function of peroxide wt%. We obtained the experimental data in Figure 5 by measuring the velocity of approximately 200 nanomotors using optical microscopy in varying concentrations of hydrogen peroxide. The value of j/j_d at each peroxide concentration is estimated from the published dependence of electrocatalytically generated current density on Pt and Au interdigitated microelectrodes [34]. We have subtracted out the characteristic Brownian velocity of the nanomotors (measured here to be 4.87 μm s^{-1}) from all experimental data points in order to only consider the axial velocities measured in the experiments. The velocity increases linearly with peroxide concentration. We also make predictions of this velocity as a function of the current density and zeta potential of the rod as will be discussed in the following section.

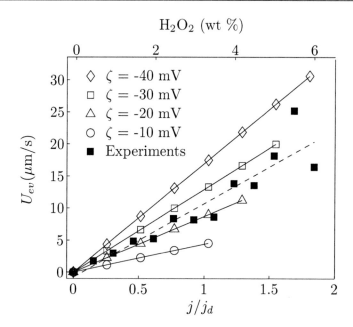

Figure 5. Nanomotor velocity as a function of dimensionless flux j/j_d (lower axis) and hydrogen peroxide concentration (upper axis). Simulations (open symbols), the scaling analysis (lines), and experiments (closed symbols) show excellent agreement. Simulations are shown for four values of the zeta potential. Copyright (2008) by the American Physical Society.

Movie 4. Nanomotors in 1% Hydrogen Peroxide

Movie 5. Nanomotors in 3% Hydrogen Peroxide. The movies show that the velocity of the motors increases with peroxide concentration.

THEORY AND SIMULATIONS

Here we present detailed simulations, scaling analysis, and experiments that describe the locomotion of bimetallic nanomotors in hydrogen peroxide solutions due to Reaction Induced Charge Auto-Electrophoresis (RICA) [29, 30]. Nanomotor movement is the result of an electroviscous slip velocity that is driven by electrical body forces resulting from charge density and electric fields that are internally generated by electrochemical reactions occurring on the particle surface. We expect that a detailed understanding of the physics underlying the nanomotors' motion will provide a basis for rational design of next-generation nanomachines capable of operation in diverse conditions and applications.

In our model, we consider a rod immersed in a binary electrolyte with equal concentrations of H^+ and OH^- ions. In the dilute solution limit, the steady ion concentration distributions are given by the advection-diffusion equation,

$$u \cdot \nabla c_i = D_i \nabla^2 c_i + z_i F \nu_i \nabla \cdot (c_i \nabla \phi), \tag{1}$$

where u is the fluid velocity, c is the molar concentration, D is the diffusivity, z is the valence, F is the Faraday constant, ν is the mobility, ϕ is the electrostatic potential, and the subscript i denotes the species. The electrostatic potential in turn depends on the local free charge density as described by the Poisson equation,

$$-\epsilon \nabla^2 \phi = \rho_e = F(z_+ c_+ + z_- c_-), \tag{2}$$

where $z_+ = 1$, $z_- = -1$, ρ_e is the volumetric charge density, and $\varepsilon = \varepsilon_r \varepsilon_0$ is the permittivity of the liquid which we assume to be constant. In our model, fluid motion (and thus nanomotor motion) is driven by electrical body forces, which depend on the concentrations of charged species only. We and others have estimated the forces due to diffusiophoresis and found that they are several orders of magnitude smaller than those produced by the induced charge mechanism described here [33]. For this reason we do not consider the oxygen and peroxide concentrations and focus on a simple binary electrolyte. To close the system of equations we include the steady, incompressible Navier-Stokes equations for a Newtonian fluid,

$$\nabla \cdot u = 0 \tag{3}$$

$$\rho(u \cdot \nabla u) = -\nabla p + \eta \nabla^2 u - \rho_e \nabla \phi. \tag{4}$$

Here ρ is the fluid density, p is the pressure, η is the dynamic viscosity, and $\rho_e \nabla \phi$ is the electrical body force that results from the coupling of charge density and electric field. This general framework has been used extensively to describe electrohydrodynamic flows [15], particularly electrokinetic flows which are driven by a coupling of externally applied fields and charged objects.

The reactions are represented by boundary conditions specifying the molar proton fluxes on the surface of the nanomotor. On the anode and cathode we prescribe equal and opposite fluxes j and $-j$ normal to the wire surface. Since anions (hydroxide ions) do not participate in the reactions, the normal anion flux is set to zero everywhere on the nanomotor surface. The values of the proton fluxes specified in the simulations are based on previously published measurements of current density at Pt and Au electrodes in hydrogen peroxide [6, 23, 34]. Here we do not directly model the electrochemical reactions that are described by the Butler-Volmer equation, because we have direct measurements of the current density for our electrocatalysts and fuel. At the nanomotor surface we apply the no-slip condition for the velocity and specify the local surface potential (relative to the bulk solution) as $\phi = \xi$. Far from the rod surface, the electrostatic potential decays to zero and ion concentrations approach their bulk value, i.e., $\phi \to 0$ and $c_i \to c_\infty$ as the radial distance $r \to \infty$.

We non-dimensionalize the momentum equation using the following scaling quantities: $|u| \propto U_{ev}$, $p \propto \eta U_{ev}/d$, $\rho_e \propto \rho_{e_0}$, and $\nabla \phi \propto E_0$ where ρ_{e_0} is a characteristic charge density, E_0 is a characteristic electric field, d is a viscous length scale, and U_{ev} is a characteristic electroviscous velocity. Applying these scalings to the momentum equation (4), a Reynolds number emerges based on the electroviscous velocity, given by $Re = \rho U_{ev} d/\eta$. Here we use the electroviscous velocity which arises in electrokinetic systems due to the balance of electrical body and viscous forces acting on the fluid, defined as [15]

$$U_{ev} \equiv \frac{\rho_{e_0} E_0}{\eta/d^2}. \tag{5}$$

Metallic nanorods support a native surface charge in aqueous solutions [9]. The charged surface attracts a screening cloud of counter-ions which develops a region of net charge density in the electrical double layer (EDL) surrounding the rod. The characteristic length scale of the charge density region is the Debye thickness λ_D, which scales with the background electrolyte concentration $c_\infty^{-1/2}$. This charge density in the EDL scales with the potential in the EDL based on the Poisson-Boltzmann equation for a symmetric binary electrolyte,

$$\rho_{e_0} \propto \frac{2z^2 F^2 c_\infty \zeta}{RT} \propto \frac{\epsilon \zeta}{\lambda_D^2}, \tag{6}$$

where $z = 1$ and we have imposed the Debye-Hückel approximation. In order to also include the effects of the reaction-driven flux, we introduce a characteristic electric field based on the flux and diffusivity of protons,

$$E_0 \propto \frac{F\lambda_D h}{\epsilon D_+} j, \tag{7}$$

where h is the length of the nanomotor and also the characteristic length for the electric field. Combining expressions (5 – 7), the electroviscous velocity scales as

$$U_{ev} \propto \frac{\zeta F h \lambda_D}{\eta D_+} j. \tag{8}$$

Here we scale the viscous length scale d with the EDL thickness λ_D since the region of significant viscous and electrical body forces is limited to the area with charge density. As shown in Figure 6B, the z-component of the self-generated electric field is significant along a distance that scales with the nanomotor length h. The scaling analysis shows that the nanomotor speed increases linearly with the reaction flux j because the flux generates charge density which produces the internally generated field. Equation 8 can be recast in the form of the Helmholtz-Smoluchowski equation $U_{ev} = \epsilon \zeta E_0/\eta$ which describes the electrophoretic velocity for a charged particle in the presence of an external electric field [14], where here E_0 is given by equation 7.

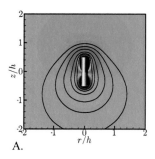 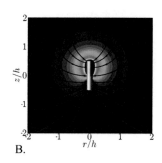 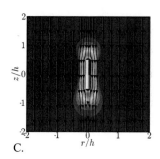

A. B. C.

Figure 6. Simulation-generated plots of **(A)** normalized proton concentration c^* (color-red and blue denote high and low proton concentrations, respectively) and electrical potential Φ^* (contour lines), **(B)** charge density ρ_e (color- red denote positive charge density above background zero levels in blue) and electric field (streamlines) and **(C)** RICA velocity magnitude (colors – blue is zero and red is maximum velocity) and streamlines (black lines) for the case $\zeta = -10$ mV and $j/j_d = 0{:}8$. The reactions lead to an asymmetry in the proton concentration such that an excess of protons builds up at the anode and protons are depleted at the cathode. The excess of protons results in positive charge density at the anode and the generation of electric field pointing from the anode to the cathode. Copyright (2010) by the American Physical Society.

We assume the flow is axisymmetric and thus solve the system over a two-dimensional cross-section of the 3-D problem. The $100\,\mu m \times 100\,\mu m$ simulation domain is discretized into approximately 181,000 triangular mesh elements. The length and diameter of the simulated nanomotor are set to $2\,\mu m$ and 370 nm, respectively. We solve the system of governing equations $(1-4)$ numerically using the linear system solver PARDISO. Using a two-ion model with a fixed bulk electrolyte concentration, the only free parameters in the system are the nanomotor native surface potential ζ and the surface flux j. We normalize the flux by characteristic flux based on Nernst's diffusion limited current density given by $j_d = 4D_+c_\infty/\lambda_D$ [3].

Figure 6A shows the dimensionless proton concentration $c^* = (c_+ - c_\infty)/c_\infty$ (colour) and contours of the electric potential normalized by the thermal voltage $\phi^* = zF\phi/RT$ (black lines) for the case where $\zeta = -10$ mV and $j/j_d = 0.8$. In the absence of reactions, the EDL proton concentration and electrical potential are both symmetric around the nanorod. Figure 6A shows that when reaction-driven fluxes are introduced, an asymmetry in the proton concentration is established such that an excess of protons builds up at the anode and protons are depleted at the cathode. The reactions also result in an asymmetric electrical potential profile that bulges at the cathode.

Figure 6B shows the charge density and streamlines of electric field for the same case as Figure 6A. The charge density in the diffuse layer of the anode is positive because the negatively charged surface attracts cations near the surface and the surface reactions constantly inject cations. At the cathode the deficiency of protons due to reactions and the shielding protons due to the negatively charged surface nearly counteract each other resulting in weak negative charge density. The charge density generates an electric field, as described mathematically by Poisson's equation. The electric field couples with the charge density to produce an electrical body force, $\rho_e\boldsymbol{E}$, which acts on the fluid to drive an electroviscous velocity and propel the nanomotor.

Figure 6C shows the RICA velocity magnitude (colour) and streamlines (black lines). These simulations are conducted in the reference frame of the nanomotor. Fluid flows from the anode to the cathode due to electrical body forces that result from a coupling of positive charge density and electric field tangent to the nanomotor surface. By Galilean invariance, this is equivalent to the nanomotor swimming with the anode end forward.

Figure 5 shows the nanomotor velocity as a function of the flux j/j_d obtained from simulations, the scaling analysis, and experiments. The experiments and simulations show good agreement for a native surface potential of -25 mV, which we obtain from independent measurements of the zeta potential for Au and Pt particles in aqueous solutions [9, 10]. The simulations, scaling analysis, and experiments all show that the electroviscous swimming velocity of the nanomotors increase linearly with increasing peroxide concentration or current density.

SUMMARY

Nobel Laureate Peter Mitchell originally proposed that an asymmetric ion flux across an organism's membrane could generate electric fields that drive locomotion. Although this locomotion mechanism was later rejected for some species of bacteria, modern nanofabrication tools have been harnessed to engineer bimetallic Janus particles that swim by ion fluxes generated by asymmetric electrochemical reactions. Here we have presented governing equations, scaling analyses, and numerical simulations that describe the motion of bimetallic rod-shaped motors in hydrogen peroxide solutions due to reaction-induced charge auto-electrophoresis. Our simulations show strong agreement with the scaling analysis and experiments. The analysis shows that electrokinetic locomotion results from electro-osmotic fluid slip around the nanomotor surface.

We have demonstrated the controlled motion of the nanomotors through micro-channel networks. We have shown that external magnetic field control of the motors can be used to pick-up, drag, and release micron scale colloidal cargo. Synthetic nanomachines may pave the way to integrated functional microdevices powered by autonomous transport.

ACKNOWLEDGEMENTS

The authors acknowledge Jared Burdick, Rawiwan Laocharoensuk, Kamil Salloum, Marcus Herrmann for contributions and stimulating discussions. This work was sponsored by NSF graduate fellowships to JLM and PMW and grants CBET-0853379 and CHE 0506529.

REFERENCES

[1]   Anderson, J.L. (1989) Colloid transport by interfacial forces. *Annu. Rev. Fluid Mech.* **21**:61 – 100.
        doi: 10.1146/annurev.fl.21.010189.000425.

[2]   Balasubramanian, S., Kagan, D., Manesh, K.M., Calvo-Marzal, P., Flechsig, G.-U., and Wang, J. (2009) Thermal modulation of nanomotor movement. *Small* **5**(13):1569 – 1574.
        doi: 10.1002/smll.200900023.

[3]   Bazant, M.Z., Chu, K.T., and Bayly, B.J. (2005) Current-voltage relations for electrochemical thin films. *SIAM J. Appl. Math.* **65**(5):1463 – 1484.
        doi: 10.1137/040609938.

[4]   Brennen, C. and Winet, H. (1977) Fluid mechanics of propulsion by cilia and flagella. *Annu. Rev. Fluid Mech* **9**(1):339 – 398.
        doi: 10.1146/annurev.fl.09.010177.002011.

[5]   Burdick, J., Laocharoensuk, R., Wheat, P.M., Posner, J.D. and Wang, J. (2008) Synthetic nanomotors in microchannel networks: Directional microchip motion and controlled manipulation of cargo. *J. Am. Chem. Soc.* **130**(26):8164 – 8165.
        doi: 10.1021/ja803529u.

[6]   Calvo-Marzal, P., Manian Manesh, K., Kagan, D., Balasubramanian, S., Cardona, M., Flechsig, G.-U., Posner, J. and Wang, J. (2009) Electrochemically-triggered motion of catalytic nanomotors. *Chem. Commun.* **30**:4509 – 4511.
        doi: 10.1039/b909227g.

[7]   Catchmark, J.M., Subramanian, S. and Sen, A. (2005) Directed rotational motion of microscale objects using interfacial tension gradients continually generated via catalytic reactions. *Small* **1**(2):202 – 206.
        doi: 10.1002/smll.200400061.

[8]   Córdova-Figueroa, U.M. and Brady, J.F. (2008) Osmotic propulsion: The osmotic motor. *Phys. Rev. Lett.* **100**(15).
        doi: 10.1103/PhysRevLett.100.158303.

[9]   Dougherty, G.M., Rose, K.A., Tok, J.B., Pannu, S.S., Chuang, F.Y.S., Sha, M.Y., Chakarova, G. and Penn, S.G. (2008) The zeta potential of surface-functionalized metallic nanorod particles in aqueous solution. *Electrophoresis* **29**(5):1131 – 1139.
        doi: 10.1002/elps.200700448.

[10]   Du, H.-Y., Wang, C.-H., Hsu, H.-C., Chang, S.-T., Chen, U.-S., Yen, S.C., Chen, L.C., Shih, H.-C., and Chen, K.H. (2008) Controlled platinum nanoparticles uniformly dispersed on nitrogen-doped carbon nanotubes for methanol oxidation. *Diamond and Related Materials* **17**(4 − 5):535 − 541.
doi: 10.1016/j.diamond.2008.01.116.

[11]   Gibbs, J.G. and Zhao, Y.P. (2009) Autonomously motile catalytic nanomotors by bubble propulsion. *Appl. Phys. Lett.* **94**(16).
doi: 10.1063/1.3122346.

[12]   Golestanian, R., Liverpool, T.B. and Ajdari, A. (2005) Propulsion of a molecular machine by asymmetric distribution of reaction products. *Phys. Rev. Lett.* **94**(22).
doi: 10.1103/PhysRevLett.94.220801.

[13]   Harold, F.M., Bronner, F., Slayman, C.L., and Kleinzeller, A. (1982) Pumps and currents: A biological perspective. *In* Electrogenic Ion Pumps, volume 16 of *Current Topics in Membranes and Transport*, pages 485 − 516. Academic Press.

[14]   Henry, D.C. (1931) The cataphoresis of suspended particles. part i. the equation of cataphoresis. *Proceedings of the Royal Society of London*. Series A, Containing Papers of a Mathematical and Physical Character 133(821):106 − 129.
doi: 10.1098/rspa.1931.0133.

[15]   Hoburg, J.F. and Melcher, J.R. (1976) Internal electrohydrodynamic instability and mixing of fluids with orthogonal field and conductivity gradients. *J. Fluid Mech.* **73**(2):333 − 351.
doi: 10.1017/S0022112076001390.

[16]   Ismagilov, R.F., Schwartz, A., Bowden, N. and Whitesides, G.M. (2002) Autonomous movement and self-assembly. *Angew. Chem., Int. Ed.* **41**(4):652 − 654.
doi: 10.1002/1521-3773(20020215)41:4<652::AID-ANIE652>3.0.CO;2-U.

[17]   Jaffe, L.F. and Nuccitelli, R. (1974) Ultrasensitive vibrating probe for measuring steady extracellular currents. *J. Cell Biol.* **63**(2):614 − 628.
doi: 10.1083/jcb.63.2.614.

[18]   Kagan, D., Calvo-Marzal, P., Balasubramanian, S., Sattayasamitsathit, S., Manesh, K.M., Flechsig, G.-U. and Wang, J. (2009) Chemical sensing based on catalytic nanomotors: Motion-Based detection of trace silver. *J. Am. Chem. Soc.* **131**(34):12082.
doi: 10.1021/ja905142q.

[19]   Kline, T.R., Paxton, W.F., Mallouk, T.E. and Sen, A. (2005) Catalytic nanomotors: Remote-controlled autonomous movement of striped metallic nanorods. *Angew. Chem., Int. Ed.* **44**(5):744 − 746.
doi: 10.1002/anie.200461890.

[20]   Kline, T.R., Paxton, W.F., Wang, Y., Velegol, D., Mallouk, T.E. and Sen, A. (2005) Catalytic micropumps: Microscopic convective fluid flow and pattern formation. *J. Am. Chem. Soc.* **127**(49):17150 – 17151.
doi: 10.1021/ja056069u.

[21]   Kline, T.R., Iwata, J., Lammert, P.E., Mallouk, T.E:, Sen, A. and Velegol, D. (2006) Catalytically driven colloidal patterning and transport. *J. Phys. Chem. B* **110**(48):24513 – 24521.
doi: 10.1021/jp064393l.

[22]   Lammert, P.E., Prost, J. and Bruinsma, R. (1996) Ion drive for vesicles and cells. *J. Theor. Biol.* **178**(4):387 – 391.
doi: 10.1006/jtbi.1996.0035.

[23]   Laocharoensuk, R., Burdick, J. and Wang, J. (2008) Carbon-nanotube-induced acceleration of catalytic nanomotors. *ACS Nano* **2**(5):1069 – 1075.
doi: 10.1021/nn800154g.

[24]   Lauga, E. and Powers, T.R. (2009) The hydrodynamics of swimming microorganisms. *Rep. Prog. Phys.* **72**(9).
doi: 10.1088/0034-4885/72/9/096601.

[25]   Lund, E.J. (1947) Bioelectric fields and growth. U. Texas Press, Austin.

[26]   Mano, N. and Heller, A. (2005) Bioelectrochemical propulsion. *J. Am. Chem. Soc.* **127**(33):11574 – 11575.
doi: 10.1021/ja053937e.

[27]   Mitchell, P. (1956) Hypothetical thermokinetic and electrokinetic mechanisms of locomotion in micro-organisms. *Proc. R. Phys. Soc. Edinburgh* **25**:32 – 34.

[28]   Mitchell, P. (1972) Self-Electrophoretic locomotion in microorganisms – bacterial flagella as giant ionophores. *FEBS Letters* **28**(1):1 – 4.
doi: 10.1016/0014-5793(72)80661-6.

[29]   Moran, J.L., Wheat, P.M. and Posner, J.D: (2010) Locomotion of electrocatalytic nanomotors due to reaction induced charge Auto-Electrophoresis. *Phys. Rev. E* **81**(6):065302.
doi: 10.1103/PhysRevE.81.065302.

[30]   Moran, J.L. and Posner, J.D. Electrokinetic locomotion due to reaction induced charge electrophoresis. submitted to *J. Fluid Mech.* September 2010.

[31]   Nuccitelli, R.and Jaffe, L.F. (1976) Ionic components of current pulses generated by developing fucoid eggs. *Dev. Biol.* **49**(2):518 – 531.
doi: 10.1016/0012-1606(76)90193-7.

[32] Paxton, W.F., Kistler, K.C., Olmeda, C.C., Sen, A., St Angelo, S.K., Cao, Y.Y., Mallouk, T.E., Lammert, P.E. and Crespi, V.H. (2004) Catalytic nanomotors: Autonomous movement of striped nanorods. *J. Am. Chem. Soc.* **126**(41):13424 – 13431. doi: 10.1021/ja047697z.

[33] Paxton, W.F., Sen, A. and Mallouk, T.E. (2005) Motility of catalytic nanoparticles through self-generated forces. *Chem.–Eur. J.* **11**(22):6462 – 6470. doi: 10.1002/chem.200500167.

[34] Paxton, W.F., Baker, P.T., Kline, T.R., Wang, Y., Mallouk, T.E. and Sen, A. (2006) Catalytically induced electrokinetics for motors and micropumps. *J. Am. Chem. Soc.* **128**(46):14881 – 14888. doi: 10.1021/ja0643164.

[35] Pitta, T.P. and Berg, H.C. (1995) Self-Electrophoresis is not the mechanism for motility in swimming cyanobacteria. *J. Bacteriol.* **177**(19):5701 – 5703.

[36] Purcell, E.M. (1977) Life at low reynolds number. *Am. J. Phys.* **45**(1):3 – 11. doi: 10.1119/1.10903.

[37] Sundararajan, S., Lammert, P.E., Zudans, A.W., Crespi, V.H. and Sen, A. (2008) Catalytic motors for transport of colloidal cargo. *Nano Letters* **8**(5):1271 – 1276. doi: 10.1021/nl072275j.

[38] Wang, Y., Hernandez, R.M., Bartlett, J., Bingham, J.M., Kline, T.R., Sen, A. and Mallouk, T.E. (2006) Bipolar electrochemical mechanism for the propulsion of catalytic nanomotors in hydrogen peroxide solutions. *Langmuir* **22**(25):10451 – 10456. doi: 10.1021/la0615950.

[39] Waterbury, J.B., Willey, J.M., Franks, D.G., Valois, F.W. and Watson, S.W. (1985) A cyanobacterium capable of swimming motility. *Science* **230**(4721):74 – 76. doi: 10.1126/science.230.4721.74.

[40] Yates, G.T. (1986) How microorganisms move through water. *Am. Sci.* **74**(4):358 – 365.

[41] Peter D. Mitchell was a British biochemist who was later awarded the 1978 Nobel Prize in Chemistry for his chemiosmotic mechanism of ATP synthesis.

Control of Rotary Motion at the Nanoscale: Motility, Actuation, Self-assembly

Petr Král*, Lela Vuković, Niladri Patra, Boyang Wang and Alexey Titov

Department of Chemistry, University of Illinois at Chicago,
Chicago, IL 60607, U.S.A.

E-Mail: *pkral@uic.edu

Received: 21st September 2010 / Published: 13th June 2011

Abstract

Controlling motion of nanoscale systems is of fundamental importance
for the development of many emerging nanotechnology areas. We review a variety of mechanisms that allow controlling rotary motion in
nanoscale systems with numerous potential applications. The discussed
mechanisms control molecular motors and propellers of liquids, nanoscale objects rolling on liquids, nanochannels with rotary switching
motion and self-assembly of functional carbonaceous materials guided
by water nanodroplets and carbon nanotubes.

Introduction

Today, manipulation of nanostructures in fluids is of great importance in nano- and biotechnology applications. Many examples of successful transport of nanoscale cargo in fluids can
be found in biological systems. For example, cells can transport molecules and micelles [1]
by kinesin [2], dynein [3] or myosin [4]. After picking up the cargo, these motor proteins can
move along tubulous filaments. Other molecular motors can realize numerous types of
motions [5], such as bacteria self-propelling by tiny flagella [6]. Controlled delivery of
molecules and nanoscale objects *in vitro* can be realized by modifying existing natural

systems to perform desired transport tasks [7]. This approach has been successfully tested in microfluidic systems, where natural molecular motors were shown as good candidates for powering the transport of nanoparticles and other cargo [8 – 10].

Artificial molecular motors have been also synthesized [11]. They could be driven by optical [12 – 15], electrical [16], chemical [17], thermal [18], and other ratchet-like means [19, 20]. With the availability of nanoscale propelling units, the lab-on-a-chip concept [21] could be modified to study single molecules [22 – 24], and extended to another lab-in-a-cell concept, where individual biomolecules inside cells could be transported [25, 26].

Carbon-based materials with at least one spatial dimension at the nanoscale, such as carbon nanotubes, nanocones and graphene [27 – 29], have many unique properties that make them ideal for the fabrication of active nanofluidic elements. They are chemically stable, very strong and rigid, since covalent binding of their carbon atoms is realized by planar sp^2-orbitals. Carbon nanotubes (CNTs) can be modified by physisorption [30] or covalent bonding of molecular ligands [31]. The semi-metallic graphene could be also tuned by doping [32] and chemical functionalization [33 – 37]. The functionalized carbon nanostructures are promising building blocks of multi-component nanodevices, which could perform complex tasks. In this work, we explore potential nanotechnology and nanofluidics applications of these and other materials, including surfactant-covered metallic nanoparticles and hybrid (inorganic, organic, biomolecular) structures.

CONTROL OF MOTILITY AT THE NANOSCALE

In order to design autonomous nanoscale devices in liquids, we need to understand the physical laws which determine their motion. Flow of fluids around nanoscale structures is characterized by low Reynolds numbers (ratio of inertial and viscous forces), $R = a\nu\rho/\eta$, where a is the size of the system, ν its velocity, ρ its density and η is the viscosity of the fluid. In low-R environments, inertia is negligible, and viscosity dominates the motion of objects. According to "scallop theorem," systems can pump fluids and propel themselves in environments with low Reynolds numbers only if their motion is not time-reversible [38]. Unidirectional rotary motion is non-reversible and thus suitable for driving nanoscale processes in liquids.

Molecular propellers

Propellers are commonly used on macroscale to convert rotary motion into thrust. The performance of nanoscopic propellers with "chemically tunable" blades has been studied by molecular dynamics (MD) simulations [39]. The propellers are formed by (8,0) carbon nanotubes (CNT) [40], with chemically attached aromatic blades. These propellers could potentially be synthesized by cyclic addition reactions [41]. Two types of propellers were designed: (1) the bulk propeller, shown in Figure 1 (left), can pump liquid along the tube z-axis, due to two blades being tilted with respect to the nanotube axis; (2) the surface propeller, shown in Figure 1, pumps water orthogonal to the tube axis by four larger blades aligned straight along the axis.

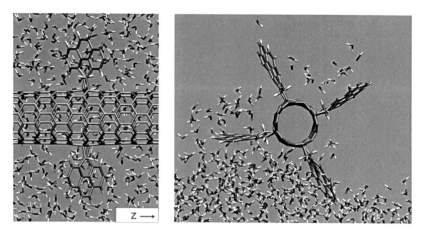

Figure 1. The bulk (left) and surface (right) water propellers that pump water along the tube (z) axis and orthogonal to it, respectively.

Movie 1. Hydrophic bulk propeller

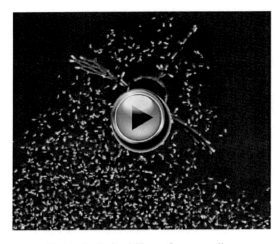

Movie 2. Hydrophilic surface propeller

Movie 3. Hydrophobic surface propeller

Simulations have shown that these propellers can pump liquids when driven by constant applied torque on the CNT. Figure 2 shows the temperature dependence of the rotation rates (left) and pumping rates (right) of the bulk hydrophobic and hydrophilic propellers with different blades, obtained in the hydrophobic dichloromethane (DCM) and hydrophilic water solvents. As the system is heated above the (normal) freezing points of the solvents, $T^{DCM}_f = 175$ K and $T^{water}_f = 273$ K, the rotation rates grow, due to smaller solvent viscosities. The hydrophilic propeller rotates slower, since its polar blades interact stronger with both solvents. Substantially slower rotation occurs in water that forms hydrogen bonds with its polar blades [42] (see inset in Figure 2).

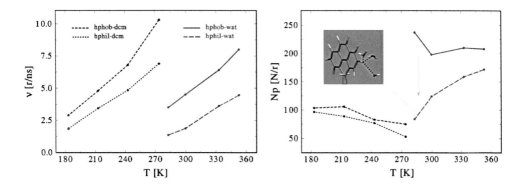

Figure 2. (left) Rotation rates (round/ns) of the bulk hydrophobic ("pho") and hydrophilic ("phi") propellers in water and DCM solvents as a function of temperature. (right) The pumping rates (molecules/round) of the bulk hydrophobic and hydrophilic propellers in water and DCM solvents as a function of temperature. (inset) Formation of hydrogen bonds between the hydrophilic blades and water can dramatically reduce the pumping rate.

The pumping rates are significantly different in two solvents, as seen in Figure 2 (right). In the DCM solvent, the pumping rates are low due to the weak coupling between the solvent molecules and the blades. Solvent molecules resemble ideal gas that "slips" on the blades in independent scattering events. The same is true for both hydrophobic and hydrophilic propellers, although in the hydrophilic case weak Coulombic coupling of the blades and DCM slightly slows down the pumping. At higher temperatures, momentum is dissipated faster and the pumping rate is reduced. In the polar (water) solvent, water molecules form clusters, around the blades, transiently held together by hydrogen bonds. This can effectively increase the cross section of the hydrophobic blades and the pumping rate, in agreement with Figure 2 (right). In the case of the hydrophilic propeller, water forms relatively stable hydration shells around the blades that reduce the effective space available for a direct contact of the pumped molecules with the blades. This causes drastically smaller pumping rates with respect to the hydrophobic propeller. At high temperatures, the hydrogen bonds break down and the pumping rates increase and tend to be the same in both types of propellers.

These molecular propellers could be used as functional components of nanomachines. One can imagine that a nanomotor (the gamma subunit of the F_1-ATPase enzyme), driven by the ATP hydrolysis, can be linked to the molecular propeller described above. In this way, a functional unit capable of transporting nanoscale cargo *in vitro* and *in vivo* could be obtained.

Molecular motors

Rotary motors could drive the above described molecular propellers. At the macroscale, electronic driving is the most efficient method of powering motors [43]. At the nanoscale, electronic driving could be achieved by *electron tunnelling*, which can induce periodic vibrational [44] and translational motion in molecules [45]. Here, we show how nanoscale vibrations can be effectively transformed into concerted rotary motion by electron tunnelling [46].

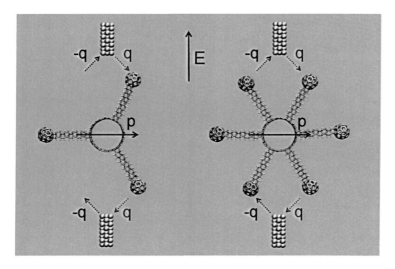

Figure 3. Tunneling-driven nanoscale motors with three (left) and six (right) fullerene blades. In an external homogeneous electric field E oriented along the vertical z direction, the electron tunneling from the neutral electrodes to the blades maintains an electric dipole p on the rotor, which is on average orthogonal to the field direction. This dipole is unidirectionally rotated by the electric field.

In Figure 3, we present two types of tunnelling-driven molecular motors, made with three (left) and six (right) stalks [46]. Their shaft is formed by a (12,0) carbon nanotube (CNT) [41], which could be mounted into CNT bearings [47]. The stalks, formed by polymerized iceane molecules with saturated bonds [48], are attached to the shaft at an angle of 120° or 60° with respect to each other. The length of stalks was chosen to prevent nonresonant electron tunnelling from the blades to the shaft [49]. The energies of their electronic states should also prevent the electron transfer along the stalks by resonant tunnelling [50]. The blades are made of molecules with conjugated bonds (fullerenes) covalently attached at the top of the stalks. A similar single molecule rotary motor based on a ruthenium complex was recently synthesized [51]. In our MD simulations, an external homogeneous electrostatic field ε, oriented along the vertical z direction, has been used for periodical charging and discharging of the blades by electron tunnelling from two neutral electrodes, placed and immobilized in the proximity of the motor, as shown in the Figure 3. The field ε then powers the system by rotating the formed *dipole p* of the rotor that is on average orthogonal to the field direction.

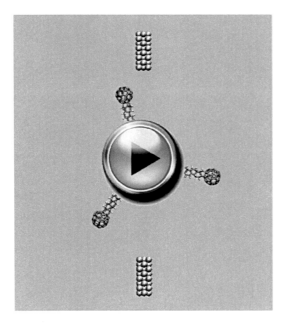

Movie 4. Rotary tunneling motor with 3 blades

In Figure 4, we show the efficiency of these tunnelling-driven motors obtained by applying a damping torque and stabilizing the motor rotation to the steady-state. In the limit of high damping torque (high loading), the efficiency saturates to the values of $\eta \approx 0.85$ and $\eta \approx 0.6$ for three- and six-fullerene motors, respectively. The maximum efficiency of the motors, $\eta = 1$, is never observed due to the dissipation of the torque to the internal damping phonon modes of the motor. For lower damping torques (low loading), the efficiency of these motors is vanishing, $\eta \rightarrow 0$, similar to the macroscopic electric motors [43]. These simulations demonstrate that rotary synthetic molecular motors can have robust performance under load, and in the presence of noise and defects.

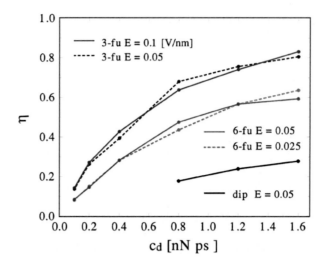

Figure 4. Dependence of the efficiency η for the 3-fullerene and 6-fullerene motors on the damping coefficient c_d. At larger c_d the efficiency grows and becomes stabilized, but $\eta_{3-fu} > \eta_{6-fu}$.

Molecular rollers

One can move nanoscale objects by AFM, light tweezers [52, 53] and other microscopic techniques. However, future nanosystems might need to move autonomously. Electrically, magnetically or optically active objects could be set in motion by the application of external fields. Nanoobjects responsive to external stimuli can be designed by appropriate chemistry.

Recently, this idea has been explored [54] by means of atomistic MD simulations, where surfactant-covered nanorods were rolled (and translated) on water surface when driven electrically. In Figure 5, we show a model of the alkane-covered nanorod, which has the empty coarse grained core that is locally and transiently charged on its surface. The driving of the model nanorods was simulated by adding small rechargeable electrodes on their surfaces: a narrow stripe of beads, positioned in the layer adjacent to the core and parallel to the water surface, is sequentially recharged (see Figure 5). We test two recharging regimes, where one stripe along the roller circumference is always homogeneously charged with one electron. In the "quasi-stochastic" regime, recharging of the stripe shifted counter clockwise by $\theta = \pi/6$ radians from the actually charged stripe is performed once its angle with the water surface drops below $\pi/6$ radians. In the "periodic" regime, recharging of the next counter clockwise positioned stripe is realized periodically every t_C, giving a charged wave on the surface.

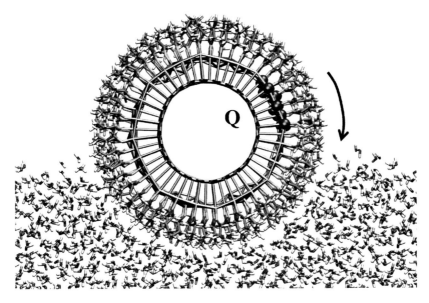

Figure 5. Model nanorod rolling on the water surface by Coulombically induced torque.

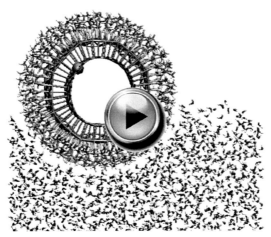

Movie 5. Rolling of nanorods on water by light

In reality, metallic nanorods can be covered with triblock surfactants, with the middle blocks made of partially oriented and photoactive chromophores, whereas inner and outer blocks could be inert and hydrophobic alkane chains. These nanorods could be rotated by excitation of their chromophores with a light beam tilted with respect to the water surface. Coulombic attraction of the transiently and asymmetrically polarized or charged chromophores in these nanorods to the polar water molecules should cause unidirectional reorientation (rolling) of the nanorods.

We have shown that with the proposed driving, nanorods with radii of $R_N = 2$- 5 nm can roll on water with translational velocities of $v_{tran} = 1$- 5 nm/ns [54]. The extent of coupling between rotational and translational motions is given by the efficiency of the rolling motion, defined as $S = v_{tran}/R_N \, v_{rot}$, where v_{rot} is the angular velocity. It turns out that translational velocities and efficiencies of the rolling motion depend on the coupling strength between surfactants and water, ε_s. While in the ideal (non-slipping) system, $S \rightarrow 1$, surfactant-covered nanorods roll on the dynamic water surface with greatly reduced efficiencies of $S = 0.1$ - 0.4. These simulations demonstrate that nanorods need to couple well to the water surface but not solvate in order to become propelled with minimal slipping during rotation [55].

ROTARY ACTUATORS IN NANOFLUIDICS

In order to control molecular flows in nanofluidics, novel synthetic nanopores might be also designed. Such pores could be based on principles similar to those operating in channel proteins. The protein cores are formed by precisely arranged sequences of amino acids that can efficiently recognize and guide the passing molecular species. Synthetic systems can have simpler structures, based on artificial proteins [55 – 58], coenzymes [59], and inorganic materials, such as zeolites [60], carbon [61] or silica [62]. Below, we describe recently explored molecular nanochannels, where rotary motion can control their selectivity and passage rates.

Molecular nanovalve

Previously, we have designed model hybrid nanochannels formed by covalently connected carbon nanocones [63]. In Figure 6 (left), we show two carbon nanocones that are stacked and connected by aliphatic chains at their open tips, in analogy to aquaporins. The functionality of this nanovalve structure is based on the fact that the size of the nanopore is altered when one nanocone is gradually rotated with respect to the other. If we mutually rotate them by a certain angle θ_{rot}, the aliphatic chains become stretched, helically wrapped, and entangled inside the nanopore. The rotation of nanocones is precise and reproducible, due to the π-π stacking.

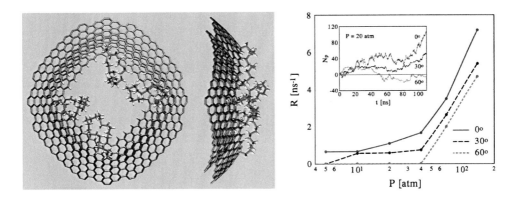

Figure 6. (left) The stacked (untwisted) pair of nanocones. (right) Dependence of the rate of pentane flow through the nanopore on the applied pressures, at rotation angles of $\theta_{rot} = 0°$, $30°$ and $60°$. (inset) The time dependence of total number of pentane molecules that pass the nanopore N_p, at the applied pressure of $P = 20$ atm and different θ_{rot}.

The size of the nanopore controls the flow of fluids through the nanovalve. In Figure 6 (right), we plot the dependence of the flow rates R on the pressure P at $\theta_{rot} = 0°$, $30°$ and $60°$. It turns out that R depends dramatically on both parameters. At $\theta_{rot} = 0°$, the cone is opened and R is nonzero even at small pressures. At large pressures, R grows and is proportional to the pressure. At the angles where the cone is largely closed, $\theta_{rot} = 30°$, $60°$, the flow is almost zero at low pressures, $P < 10$ atm. At higher pressures, the cross section area of the nanopore can become slightly bigger for $\theta_{rot} = 30°$, resulting in small constant flow in the region of $P = 10 - 40$ atm. At $P > 40$ atm, the four alkyl chains start to open for all the studied angles, and the flow increases with pressure.

In the inset of Figure 6 (right), we also show the time dependence of the total number of pentanes that pass the nanopore at $P = 20$ atm. The size of the time-dependent fluctuations in N_p reflects fluctuations in the sizes of the nanopore, caused by the alkyl chains that rotate and change their conformations. For more closed pores, the fluctuations of N_p are smaller, due to stretched alkyl chains.

Biomimetic ammonia switch

Nanochannels for selective molecular passage can be designed in direct analogy to channel proteins. Recently, we explored hybrid nanochannels with inorganic and polypeptide components by means of MD simulations [63]. In Figure 7 (left), we show a nanochannel formed by two carbon nanotubes joined by a cylindrical structure of antiparallel peptide chains. Rotation of one nanotube end with respect to the other results in either forward or backward-twisted peptide barrels in the middle of the nanochannel.

In Figure 7 (right), we present the hysteresis curve for the dependence of the twist angle on the torque applied to the parallel 8-residue structure. The results are obtained in one twisting trajectory going between the two end points (wrapped cylinders) and back, where each data point is averaged over 1,000 frames separated by 100 fs, after equilibration for 1 ns. The hysteresis curve shows that the structure can be stabilized in two configurations with non-zero twist angles, upon removal of the torque.

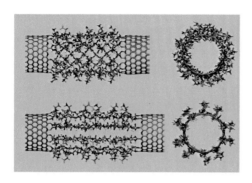 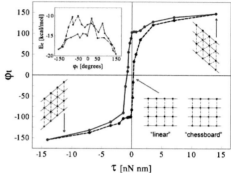

Figure 7. (left) Two types of 12-residue hybrid nanochannels solvated in water at T = 300 K. (right) The twist angle-torque hysteresis curve obtained in one cycle for the parallel 8-residue structure. Additional schemes show the charges and their mutual coupling in the straight and twisted structures. (inset) The related Coulombic energy-twist angle hysteresis curve.

These hybrid nanochannels are highly tunable due to a large selection of amino acids which can coat the inner channel walls. By screening the amino acid residues inside the peptide barrel, we identified those which make the channel interior selective to passage of ammonia. In Figure 8 (left), we show the interiors of this channel containing two tryptophan (TRP) and two tyrosine (TYR) residues for opposite twists, giving different internal structures. In Figure 8 (right), we plot the number N_p of NH_3 molecules passed under steady-state conditions through these channels within the simulation time t. The results show dramatically different flow rates, $R_{NH_3} = N_p/t$, for the forward and backward twisted configurations of the channel, $R^{for}_{NH_3} \approx 0.18$ ns^{-1}, $R^{bac}_{NH_3} \approx 0$, respectively. In the forward twist, two pairs of interior residues shadow each other, thus increasing the cross section area available for the NH_3 passage. In the backward twist, the TRP and TYR residues are spread around the circumference of the channel, thus reducing the cross section area available for the NH_3 passage. These results strongly demonstrate that this highly tunable barrel junction can act as a mechanical switch for fluid flow.

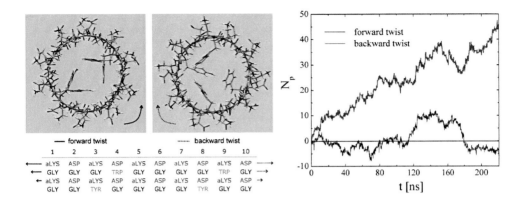

Figure 8. (left) Axial view of the two twisted configurations of the parallel 4-residue 10-peptide junction, with 2 tryptophan and 2 tyrosine residues in its interior. (left-bottom) Amino acid sequence map of 4-residue 10-peptide junction. Each column corresponds to a peptide chain, and the odd/even rows are residues with sidechains that are exposed outwards/inwards. (right) The time dependence of the total number N_p of NH_3 molecules passed through the twisted nanochannels.

Graphene nanopores

Selective molecular passage can also be realized in inorganic materials. For example, porous graphene monolayers can serve as membranes which selectively pass molecules without mechanical action. Recently, we modelled the passage of ions through functionalized nanopores in graphene monolayers by MD simulations [64]. In Figure 9, we display the studied pores with diameter of ≈ 5 Å, terminated by either negatively charged nitrogens and fluorines (left) or by positively charged hydrogens (right).

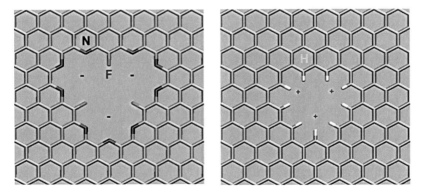

Figure 9. Functionalized graphene nanopores. (left) The F–N-terminated nanopore. (right) The H-terminated nanopore.

In Figure 10 (right insets), we show the configurations of the Na^+ and Cl^- ions passing through the two nanopores. The polar and charged nanopore rim can replace several water molecules from the first hydration shell of the ion passing through it. Therefore, the passing ion is surrounded by two separated first hydration "half-shells" at both sides. The close contact between the ion and the pore rim allows very efficient mapping and eventual removal of the ion shells, leading to the high selectivity of the ion passage.

Figure 10 (left insets) displays the time-dependent distance d between the ions and their pore centres, obtained at a low field of $E = 6.25$ mV/nm. The distance fluctuations reveal very different dynamics in two cases. The Na^+ ion passes the F–N-pore fast, without significantly binding with it. The ion rarely gets closer than 5 Å to the pore centre and stays most of the time in the water region (d > 10 Å). The ion passes through one of the three smaller holes in the nanopore with the C^3 symmetry. These asymmetric holes can not easily break the hydration shell of the Na^+ ion, which prohibits its prolonged stay in the pore. Simulations of K^+ ion show that its hydration shell can be more easily replaced by the pore rim upon shell partial removal, due to its weaker binding to the hydration shell.

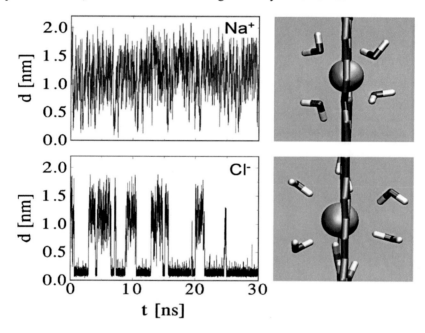

Figure 10. (left) Time-dependent distance d between the Na^+ and Cl^- ions and the centers of the F–N-pore and H-pore, respectively, at the field of $E = 6.25$ mV/nm. The dynamics of passage of these ions through the two pores is very different. (right) While both ions are surrounded by two water half-shells when passing through their pores, only the Cl^- ion has relatively stable binding to the H–pore.

Movie 6. Passage of Na^+ ion through F-N-pore in graphene

Movie 7. Passage of Cl^- ion through H-pore in graphene

The Cl^- ion has even more stable binding to the symmetric H-pore, where it stays for $\approx 70\%$ of the time. At weak fields, the Cl^- ion enters and leaves the H-pore with almost the same rates from both sides. The electric field decreases the ion-pore binding, which is reflected in shorter time periods spent by the ion in the pore. At a field of $E = 0.1$ V/nm, the Cl^- ion stays in total only 30% of the time inside the pore. These and other recent theoretical and experimental studies [7, 65 – 67] indicate that porous graphene could serve as a sieve of high selectivity and transparency, and may be sensitive enough for applications including DNA sequencing.

SELF-ASSEMBLY OF GRAPHENE-BASED NANOSTRUCTURES

We may also try to control molecular motion during self-assembly into materials with unique shapes, interactions and properties. Carbon-based materials, including graphene, carbon nanotubes (CNTs), and their functionalized forms, can form such promising materials. Currently, graphene is intensively studied for its numerous potential applications [68 – 77]. For example, graphene nanoribbons (GNRs) have been synthesized [14, 78 – 80], and etched by using lithography [81, 82] and catalytic methods [83 – 85]. Graphene flakes with strong interlayer vdW coupling and chemical functionalization at their edges can self-assemble into larger structures [86 – 88]. Individual flakes with high elasticity [89 – 91] could fold into a variety of 3D structures [92], such as carbon nanoscrolls [93 – 96].

Droplet-driven self-assembly of graphene

Recently, we have studied how graphene stripes of different shapes can roll on themselves or around other structures (water droplets or CNTs) and form novel materials [97]. We have shown that water nanodroplets can induce rapid bending, folding, sliding, rolling, and zipping of planar nanostructures, which can lead to the assembly of nanoscale sandwiches, capsules, knots, and rings. In Figure 11 (top-left), we show that once a nanodroplet with $N_w = 1,300$ water molecules is placed at T = 300 K above a star-shaped flake, the droplet binds by vdW coupling to it and induces its bending and closing within t ≈ 1 ns.

We have also shown that NDs can induce folding of graphene nanoribbons. As shown in Figure 11 (top right), when a ND ($N_w = 1,300$) is positioned above the free end of the ribbon (30×2 nm^2), the free end folds into a knot structure, touches the ribbon surface, and starts to slide fast on it due to strong vdW binding. When the radius of water droplet is several times bigger than the graphene width, it forms a multilayer ring structure, similar to multi-wall carbon nanotube, as seen in Figure 11 (bottom left). We have observed that GNRs can self-assemble into different nanostructures, and explored their dependence on the GNR width (w) and the water droplet radius (R_d). When $R_d > w$ the nanodroplet can activate and guide better the GNR assembly. The results are summarized in the phase diagram shown in Figure 11 (bottom right). Experimentally, the droplets could be deposited by Dip-Pen nanolithography [98] or AFM [99].

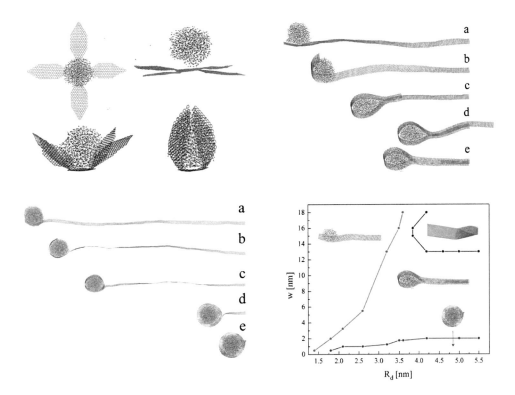

Figure 11. (top left) Nanodroplet-assisted folding of a star-shaped graphene flake. (top right) Folding and sliding of a graphene ribbon with the size of 30×2 nm^2, which is activated and guided by a nanodroplet with $N_w = 1,300$ waters and the radius of $R_d \approx 2.1$ nm. (bottom left) Folding and rolling of a graphene ribbon with the size of 90×2 nm^2, which is activated and guided by a nanodroplet with Nw = 10,000 waters and the radius of $R_d \approx 4.2$ nm. **(a-b)** The ribbon tip folds around the water droplet into a wrapped cylinder, and **(c-e)** the wrapped cylinder is induced to roll on the ribbon surface **(c-e)**. (bottom right) The phase diagram of graphene nanoribbon folding by water nanodroplets. We can see the nonfolding, sliding, rolling, and zipping phases.

Movie 8. Self-assembly of graphene flake

Movie 9. Sliding self-assembly of graphene stripe

Nanotube-driven self-assembly of graphene

We have also tested if CNTs and other nanoscale materials can activate and guide the self-assembly of GNRs on their surfaces and in their interiors [100]. In Figure 12, we present the results of our simulations of the CNT-assisted GNR self-assembly, sone at T = 300 K. Figure 12 (a-c) shows the self-assembly of the GNR (40 × 3 nm^2) once placed on the surface of (60,0) CNT (with two ends fixed). Figure 12 (d-e) shows the self-assembly of this GNR inside the same CNT. Figure 12 (f) shows that two GNRs (40 × 1 nm^2) can self-assemble into a double helix inside the same CNT. These results illustrate that GNR self-assembly can be activated and guided by CNTs.

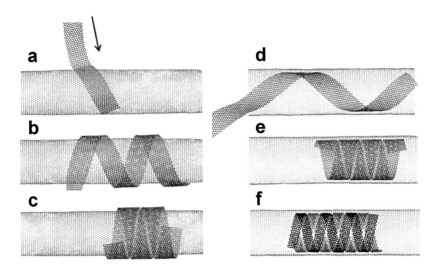

Figure 12. (a-c) Folding and rolling of a GNR with the size of 40×2 nm^2, when it is placed at the angle of 60 ° with respect to the axis of the (60,0) CNT. **(d-e)** Folding and rolling of this GNR when it is placed inside this CNT (front part of the CNT is removed for better visualisation). **(f)** Folding and rolling of two GNRs (40×1 nm^2) placed inside this CNT.

In the above described manner, one could in principle assemble arbitrary planar graphene nanostructures on the surfaces or in the interior of CNTs [101]. We test this idea on the self-assembly of a "hair-brush" like graphene nanostructure with four GNRs (60×2 nm^2). One end of all four GNRs is connected to the remaining part (40×100 nm^2) of the graphene sheet, and the distance between the adjacent GNRs is 5 nm, as shown in Figure 13 (a). Initially, the tips of all four GNRs are placed on the (20,0) CNT. They start to fold around it, and after $t \approx 4$ ns, they form a single layer ring structures on the CNT. Within other $t \approx 3$ ns, the GNRs make a multilayered ring structure. The rest of the graphene sheet also folds on the multilayered ring structures, and eventually completely wraps around the GNRs, as shown in Figure 13 (b-c). Then, we equilibrate the self-assembled structure, remove the CNT from its core, and equilibrate the remaining folded graphene. Upon removal, the diameter of the four rings slightly expands by 0.2 nm, as shown in Figure 13 (d-e), but the system retains the same shape, stabilized by the top graphene monolayer.

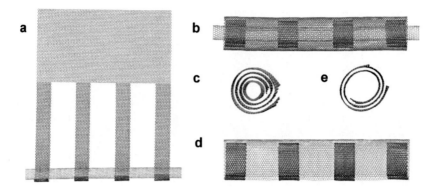

Figure 13. Graphene rings and knot formations from structured graphene flakes self-assembled on the CNT surface. **(a-c)** The formation of multiple GNR rings, covered with a single layer graphene sheet, on the (20,0) CNT. **(d-e)** After removal of the CNT, the multi-ring structure, covered with a single layer graphene sheet, is stabilized.

CONCLUSION

In summary, we have shown that control of rotary motion can be used in the preparation and activation of functional nanosystems. We discussed implementation of these principles in molecular propellers, motors, and rolling wheels. We have also shown how inorganic and biomimetic channels can be designed to serve as molecular switches, where mutual rotation of the individual components can control or selectively block the flow of different molecules. Finally, we have demonstrated that one could control the self-assembly of graphene nanostructures realized with the help of nanodroplets and graphene nanotubes. The studies can lead to the construction of building blocks in functional nanodevices, with unique mechanical, electrical, or optical properties, finding applications in electronics, sensing, and medicine.

REFERENCES

[1] Bru, R., Sanchez-Ferrer, A. and Garcia-Carmona, F. (1995) Kinetic models in reverse micelles. *Biochem. J.* **310**:721 – 739.

[2] Valle, R. D. and Milligan, R.A. (2000) The way things move: looking under the hood of molecular motor proteins. *Science* **288**:88 – 95.
doi: 10.1126/science.288.5463.88.

[3] Gennerich, A. and Valle, R.D. (2009) Walking the walk: how kinesin and dynein coordinate their steps. *Curr. Opin. Cell Biol.* **21**:59 – 67 .
doi: 10.1016/j.ceb.2008.12.002.

[4]   Bergs, J.S., Powell, B.C. and Cheney, R.E. (2001) A millenial myocin census. *Mol. Biol. Cell* **12**:780 – 794.

[5]   Schliwa, M. (Ed.) Molecular Motors. 1st ed 2002. Wiley-VCH Weinheim, Germany.

[6]   Atsumi, T., McCarter, L. and Imae, T. (1992) Polar and lateral flagellar motors of marine Vibrio are driven by different ion-motive forces. *Nature* **355**:182 – 184. doi: 10.1038/355182a0.

[7]   van den Heuvel, M.G.L. and Dekker, C. (2007) Motor proteins at work for nano-technology. *Science* **317**:333 – 336. doi: 10.1126/science.1139570.

[8]   Soong, R.K., Bachand, G.D., Neves, H.P., Olkhovets, A.G., Craighead, H.G. and Montemagno, C.D. (2000) Powering an Inorganic Nanodevice with a Biomolecular Motor. *Science* **290**:1555 – 1558.

[9]   Brunner, C., Wahnes, C. and Vogel, V. (2007) Cargo pick-up from engineered load-ing stations by kinesin driven molecular shuttles. *Lab on a Chip* **7**:1263. doi: 10.1039/b707301a.

[10]   Goel, A. and Vogel, V. (2008) Harnessing biological motors to engineer systems for nanoscale transport and assembly. *Nat. Nanotechnol.* **3**:465 – 475. doi: 10.1038/nnano.2008.190.

[11]   Green, J.E., Choi, J.W., Boukai, A., Bunimovich, Y., Johnston-Halperin, E., DeIon-no, E., Luo, Y., Sheriff, B.A., Xu, K., Shin, Y.S., Tseng, H.-R., Stoddart, J.F. and Heath, J.R. (2007) A 160-kilobit molecular electronic memory patterned at 10^{11} bits per square centimeter. *Nature* **445**:414 – 417. doi: 10.1038/nature05462.

[12]   Král, P. and Sadeghpour, H.R. (2002) Laser spinning of nanotubes: A path to fast-rotating microdevices. *Phys. Rev. B* **65**:161401 – 4(R). doi: 10.1103/PhysRevB.65.161401.

[13]   Plewa, J., Tanner, E., Mueth, D.M. and Grier, D.G. (2004) Processing carbon nano-tubes with holographic optical tweezers. *Opt. Express* **12**:1978 – 1981. doi: 10.1364/OPEX.12.001978.

[14]   Tan, S., Lopez, H.A., Cai, C.W. and Zhang, Y. (2004) Optical Trapping of Single-Walled Carbon Nanotubes. *Nano Lett.* **4**:1415 – 1419. doi: 10.1021/nl049347g.

[15]   Vacek, J. and Michl, J. (2001) Molecular dynamics of a grid-mounted molecular dipolar rotor in a rotating electric field. *Proc. Natl. Acad. Sci. U.S.A.* **98**:5481 – 5486. doi: 10.1073/pnas.091100598.

[16]   Král, P. and Seideman, T. (2005) Current-induced rotation of helical molecular wires. *J. Chem. Phys.* **123**:184702 – 184706.
doi: 10.1063/1.2107527.

[17]   Kelly, T.R., Silva, H. De and Silva, R.A. (1999) Unidirectional rotary motion in a molecular system. *Nature* **401**:150 – 152.
doi: 10.1038/43639.

[18]   Barreiro, A., Rurali, R., Hernández, E.R., Moser, J., Pichler, T., Forro L. and Bachtold, A. (2008) Subnanometer motion of cargoes driven by thermal gradients along carbon nanotubes. *Science* **320**:775 – 778.
doi: 10.1126/science.1155559.

[19]   Astumian, R.D. (1997) Thermodynamics and kinetics of a Brownian motor. *Science* **276**:917 – 922.
doi: 10.1126/science.276.5314.917.

[20]   Bull, J.L., Hunt, A.J. and Meyhofer, E. (2005) A theoretical model of a molecular-motor-powered pump. *Biomed. Microdev.* **7**:21 – 33.
doi: 10.1007/s10544-005-6168-6.

[21]   Andersson, H. and van den Berg, A. (2005) From lab-on-a-chip to lab-in-a-cell. *Proc. SPIE* 5718, 1.
doi: 10.1117/12.601553.

[22]   Craighead, H.G. (2006) Future lab-on-a-chip technologies for interrogating individual molecules. *Nature* **442**:387 – 393.
doi: 10.1038/nature05061.

[23]   Mannion, J.T. and Craighead, H.G. (2007) Nanofluidic structures for single biomolecule fluorescent detection. *Biopolymers* **85**:131 – 143.
doi: 10.1002/bip.20629.

[24]   Kellermayer, M.S.Z. (2005) Visualizing and manipulating individual protein molecules. *Physiol. Meas.* **26**:R119-R153.
doi: 10.1088/0967-3334/26/4/R02.

[25]   Strick,T., Allemand, J.-F., Croquette, V. and Bensimon, D. (2001) The manipulation of single biomolecules. *Physics Today* **54**:46 – 51.
doi: 10.1063/1.1420553.

[26]   Ying, L. (2007) Single molecule biology: Coming of age. *Mol. BioSyst.* **3**:377 – 380.
doi: 10.1039/b702845 h.

[27]   Iijima, S. (1991) Helical microtubules of graphitic carbon. *Nature* **354**:56 – 58.
doi: 10.1038/354056a0.

[28] Dresselhaus, M.S., Dresselhaus, G. and Eklund, P.C. (1996) Science of Fullerenes and Carbon Nanotubes. Academic Press Inc., San Diego.

[29] Novoselov, K.S., Geim, A.K., Morozov, S.V., Jiang, D., Zhang, Y., Dubonos, S.V., Grigorieva, I.V. and Firsov, A.A. (2004) Electric field effect in atomically thin carbon films. *Science* **306:**666 – 369.
doi: 10.1126/science.1102896.

[30] Zheng, M., Jagota, A., Strano, M.S., Santos, A.P., Barone, P., Chou, S.G., Diner, B.A., Dresselhaus, M.S., Mclean, R.S., Onoa, G.B., Samsonidze, G.G., Semke, E.D., Usrey, M. and Walls, D.J. (2003) Structure-based carbon nanotube sorting by sequence-dependent DNA assembly. *Science* **302:**1545 – 1548.
doi: 10.1126/science.1091911.

[31] Banerjee, S., Hemraj-Benny, T. and Wong, S.S. (2005) Covalent surface chemistry of single-walled carbon nanotubes. *Adv. Mater.* **17:**17 – 29.
doi: 10.1002/adma.200401340.

[32] Stankovich, S., Dikin, D.A., Dommett, G.H.B., Kohlhaas, K.M., Zimney, E.J., Stach, E.A., Piner, R.D., Nguyen, S.T. and Ruoff, R.S. (2006) Graphene-based composite material. *Nature* **442:**282 – 286.
doi: 10.1038/nature04969.

[33] Cervantes-Sodi, F., Csanyi, G., Piscanec, S. and Ferrari, A.C. (2008) Edge-functionalized and substitutionally doped graphene nanoribbons: Electronic and spin properties. *Phys. Rev. B* **77**: 165427 – 165439.
doi: 10.1103/PhysRevB.77.165427.

[34] Boukhvalov, D.W. and Katsnelson, M.I. (2008) Chemical functionalization of graphene with defects. *Nano Lett.* **8:**4373 – 4379.
doi: 10.1021/nl802234n.

[35] Ferro, S. and De Battisti, A. (2003) The 5-V window of polarizability of fluorinated diamond eletrodes in aqueous solutions. *Anal. Chem.* **75:**7040 – 7042.
doi: 10.1021/ac034717r.

[36] Dikin, D.A., Stankovich, S., Zimney, E.J., Piner, R.D., Dommett, G.H.B., Evmenenko, G., Nguyen S.T. and Ruoff, R.S. (2007) Preparation and characterization of graphene oxide paper. *Nature* **448:**457 – 460.
doi: 10.1038/nature06016.

[37] Cai, W., Piner, R.D., Stadermann, F.J., Park, S., Shaibat, M.A., Ishii, Y., Yang, D., Velamakanni, A., An, S.J., Stoller, M., An, J., Chen, D. and Ruoff, R.S. (2008) Synthesis and solid-state NMR structural characterization of 13C-labeled graphite oxide. *Science* **321:**1815 – 1817.
doi: 10.1126/science.1162369.

[38] Purcell, E.M. (1997) Life at low Reynolds number. *Amer. J. Phys.* **45**:1 – 3.

[39] Wang, B. and Král, P. (2007) Chemically tunable nanoscale propellers of liquids. *Phys. Rev. Lett.* **98**:266102 – 266105.
doi: 10.1103/PhysRevLett.98.266102.

[40] Han, J., Globus, A., Jaffe, R. and Deardorff, G. (1997) Molecular dynamics simulations of carbon nanotube-based gears. *Nanotechnology* **8**:95 – 102.
doi: 10.1088/0957-4484/8/3/001.

[41] Tasis, D., Tagmatarchis, N., Bianco, A. and M. Prato, (2006) Chemistry of Carbon Nanotubes. *Chem. Rev.* **106**:1105 – 1136.
doi: 10.1021/cr050569o.

[42] Silverstein, K.A.T., Haymet, A.D.J. and Dill, K.A. (2000) The Strength of Hydrogen Bonds in Liquid Water and Around Nonpolar Solutes. *J. Am. Chem. Soc.* **122**:8037 – 8041.
doi: 10.1021/ja000459t.

[43] Auinger, H. (2001) Efficiency of electric motors under practical conditions. *Power Eng. J.* **15**:163 – 167.
doi: 10.1049/pe:20010309.

[44] Park, H., Park, J., Lim, A.K.L., Anderson, E.H., Alivisatos, A.P. and McEuen, P.L. (2000) Nanomechanical oscillations in a single-C60 transistor. *Nature* **407**:57 – 60.
doi: 10.1038/35024031.

[45] Kaun, C.C. and Seideman, T. (2005) Current-Driven Oscillations and Time-Dependent Transport in Nanojunctions. *Phys. Rev. Lett.* **94**:226801 – 226804.
doi: 10.1103/PhysRevLett.94.226801.

[46] Wang, B., Vuković L. and Král, P. (2008) Nanoscale rotary motors driven by electron tunneling. *Phys. Rev. Lett.* **101**:186808 – 186811.
doi: 10.1103/PhysRevLett.101.186808.

[47] Cumings, J. and Zettl, A. (2000) Low-Friction Nanoscale Linear Bearing Realized from Multiwall Carbon Nanotubes. *Science* **289**:602 – 604.
doi: 10.1126/science.289.5479.602.

[48] Cupas, C.A. and Hodakowski, L. (1974) Iceane. *J. Am. Chem. Soc.* **96**:4668 – 4669.
doi: 10.1021/ja00821a050.

[49] Mikkelsen, K.V. and Ratner, M.A. (1987) Electron tunnelling in solid-state electron-transfer reactions. *Chem. Rev.* **87**:113 – 153.
doi: 10.1021/cr00077a007.

[50] Král, P. (1997) Nonequilibrium linked cluster expansion for steady-state quantum transport. *Phys. Rev. B* **56**:7293 – 7303.
doi: 10.1103/PhysRevB.56.7293.

[51] Carella, A., Rapenne, G., Launay, J.-P. (2005) Design and synthesis of the active part of a potential molecular motor. *New J. Chem.* **29**:288 – 290.
doi: 10.1039/b415214j.

[52] Ashkin, A., Dziedzic, J.M., Bjorkholm, J.E. and Chu, S. (1986) Observation of a single-beam gradient force optical trap for dielectric particles. *Opt. Lett.* **11**:288 – 290.
doi: 10.1364/OL.11.000288.

[53] Holmlin, R.E., Schiavoni, M., Chen, C.Y., Smith, S.P., Prentiss, M.G. and White-sides, G.M. (2000) Light-Driven Microfabrication: Assembly of Multicomponent, 3D Structures Using Optical Tweezers. *Angew. Chem. Int. Ed. Engl.* **39**:3503 – 3506.
doi: 10.1002/1521-3773(20001002)39:19<3503::AID-ANIE3503>3.0.CO;2-M.

[54] Vuković, L. and Král, P. (2009) Coulombically Driven Rolling of Nanorods on Water. *Phys. Rev. Lett.* **103**:246103 – 246106.
doi: 10.1103/PhysRevLett.103.246103.

[55] Qi, Z., Sokabe, M., Donowaki, K. and Ishida, H. (1999) Structure-Function Study on a de Novo Synthetic Hydrophobic Ion Channel. *Biophys. J.* **76**:631 – 641.
doi: 10.1016/S0006-3495(99)77231-0.

[56] Kuhlman, B., Dantas, G., Ireton, G.C., Varani, G., Stoddard, B.L. and Baker, D. (2003) Design of a Novel Globular Protein Fold with Atomic-Level Accuracy. *Science*, **302**:1364 – 1368.
doi: 10.1126/science.1089427.

[57] Kaplan, J. and DeGrado, W.F. (2004) De novo design of catalytic proteins. *Proc. Natl. Acad. Sci. U.S.A.* **101**: 1566 – 11570.
doi: 10.1073/pnas.0404387101.

[58] Tsai, C.-J., Zheng, J. and Nussinov, R. (2006) Designing a Nanotube Using Naturally Occurring Protein Building Blocks. *PLoS Comput. Biol.* **2**:e42.
doi: 10.1371/journal.pcbi.0020042.

[59] Murakami, Y., Kikuchi, J., Hisaeda, Y. and Hayashida, O. (1996) Artificial Enzymes. *Chem. Rev.* **96**:721 – 758.
doi: 10.1021/cr9403704.

[60] Jordan, E., Bell, R.G., Wilmer, D. and Koller, H. (2006) Anion-Promoted Cation Motion and Conduction in Zeolites. *J. Am. Chem. Soc.* **128**: 558 – 567.
doi: 10.1021/ja0551887.

[61]   Saufi, S.M. and Ismail, A.F. (2004) Fabrication of carbon membranes for gas separationa review. *Carbon* **42**:241 – 259.
doi: 10.1016/j.carbon.2003.10.022.

[62]   Duke, M.C., da Costa, J.C.D., Do, D.D.; Gray, P.G. and Lu, G.Q. (2006) Hydrothermally Robust Molecular Sieve Silica for Wet Gas Separation. *Adv. Funct. Mater.* **16**:1215 – 1220.
doi: 10.1002/adfm.200500456.

[63]   Titov, A.V., Wang, B., Sint, K. and Král, P. (2010) Controllable Synthetic Molecular Channels: Biomimetic Ammonia Switch. *J. Phys. Chem. B*, **114**:1174 – 1179.
doi: 10.1021/jp9103933.

[64]   Sint, K., Wang, B. and Král, P. (2008) Selective Ion Passage through Functionalized Graphene Nanopores. *J. Am. Chem. Soc.* **130**:16448 – 16449.
doi: 10.1021/ja804409f.

[65]   Garaj, S., Hubbard, W., Reina, A., Kong, J., Branton, D. and Golovchenko, J.A. (2010) Graphene as a subnanometre trans-electrode membrane. *Nature* **467**:190 – 193.
doi: 10.1038/nature09379.

[66]   Merchant, C.A., Healy, K., Wanunu, M., Ray, V., Peterman, N., Bartel, J., Fischbein, M.D., Venta, K., Luo, Z., Johnson, A.T.C, and Drndić, M. (2010) DNA Translocation through Graphene Nanopores. *Nano Lett.* **10**:2915 – 2921.
doi: 10.1021/nl101046t.

[67]   Nelson, T., Zhang, B. and Prezhdo, O.V. (2010) Detection of Nucleic Acids with Graphene Nanopores: Ab Initio Characterization of a Novel Sequencing Device. *Nano Lett.* **10**:3237 – 3242.
doi: 10.1021/nl9035934.

[68]   Novoselov, K.S., Geim, A.K., Morozov, S.V., Jiang, D., Zhang, Y., Dubonos, S.V., Grigorieva, I.V. and Firsov, A.A. (2004) Electric Field Effect in Atomically Thin Carbon Films. *Science* **306**:666 – 669.
doi: 10.1126/science.1102896.

[69]   Geim, A.K. and Novoselov, K.S. (2007) The rise of graphene. *Nat. Mater.* **6**:183 – 191.
doi: 10.1038/nmat1849.

[70]   Berner, S., Corso, M., Widmer, R., Groening, O., Laskowski, R., Blaha, P., Schwarz, K., Goriachko, A., Over, H., Gsell, S., Schreck, M., Sachdev, H., Greber, T. and Osterwalder, J. (2007) Boron Nitride Nanomesh: Functionality from a Corrugated Monolayer. *Angew. Chem. Int. Ed.* **46**:5115 – 5119.
doi: 10.1002/anie.200700234.

[71]   Laskowski, R., Blaha, P., Gallauner, T. and Schwarz, K. (2007) Single-Layer Model
       of the Hexagonal Boron Nitride Nanomesh on the Rh(111) Surface. *Phys. Rev. Lett.*
       98:106802 – 106805.
       doi: 10.1103/PhysRevLett.98.106802.

[72]   Tsoukleri, G., Parthenios, J., Papagelis, K., Jalil, R., Ferrari, A. C., Geim, A.K.,
       Novoselov, K.S. and Galiotis, C. (2009) Subjecting a Graphene Monolayer to Ten-
       sion and Compression. *Small* **5**:2397 – 2402.
       doi: 10.1002/smll.200900802.

[73]   Balog, R., Jorgensen, B., Nilsson, L., Andersen, M., Rienks, E., Bianchi, M., Fanetti,
       M., Laegsgaard, E., Baraldi, A., Lizzit, S., Sljivancanin, Z., Besenbacher, F., Ham-
       mer, B., Pedersen, T.G., Hofmann, P. and Hornekaer, L. (2010) Bandgap opening in
       graphene induced by patterned hydrogen adsorption. *Nat. Mater.* **9**:315 – 319.
       doi: 10.1038/nmat2710.

[74]   Yan, X., Cui, X., Li, B. and Li, L.-s. (2010) Large Solution-Processable Graphene
       Quantum Dots as Light Absorbers for Photovoltaics. *Nano Lett.* **10**:1869 – 1873.
       doi: 10.1021/nl101060 h.

[75]   Cohen-Karni, T., Qing, Q., Li, Q., Fang, Y. and Lieber, C. M. (2010) Graphene and
       Nanowire Transistors for Cellular Interfaces and Electrical Recording. *Nano Lett.*
       10:1098 – 1102.
       doi: 10.1021/nl1002608.

[76]   Wang, Y., Shao, Y., Matson, D.W., Li, J. and Lin, Y. (2010) Nitrogen-Doped Gra-
       phene and Its Application in Electrochemical Biosensing. *ACS Nano* **4**:1790 – 1798.
       doi: 10.1021/nn100315s.

[77]   Matyba, P., Yamaguchi, H., Eda, G., Chhowalla, M., Edman, L. and Robinson, N. D.
       (2010) Graphene and Mobile Ions: The Key to All-Plastic, Solution-Processed Light-
       Emitting Devices. *ACS Nano* **4**:637 – 642.
       doi: 10.1021/nn9018569.

[78]   Jiao, L., Zhang, L., Wang, X., Diankov, G. and Dai, H. (2009) Narrow graphene
       nanoribbons from carbon nanotubes. *Nature* **458**:877 – 880.
       doi: 10.1038/nature07919.

[79]   Shi Kam, N.W., Jessop, T.C., Wender, P.A. and Dai, H. (2004) Nanotube Molecular
       Transporters: Internalization of Carbon Nanotube-Protein Conjugates into Mamma-
       lian Cells. *J. Am. Chem. Soc.* **126**:6850 – 6851.
       doi: 10.1021/ja0486059.

[80] Kosynkin, D.V., Higginbotham, A.L., Sinitskii, A., Lomeda, J.R., Dimiev, A., Price, B.K. and Tour, J.M. (2009) Longitudinal unzipping of carbon nanotubes to form graphene nanoribbons. *Nature* **458**:872 – 876.
doi: 10.1038/nature07872.

[81] Tapaszto, L., Dobrik, G., Lambin, P. and Biro, L.P. (2008) Tailoring the atomic structure of graphene nanoribbons by scanning tunnelling microscope lithography. *Nat. Nanotechnol.* **3**:397 – 401.
doi: 10.1038/nnano.2008.149.

[82] Stampfer, C., Gttinger, J., Hellmller, S., Molitor, F., Ensslin, K. and Ihn, T. (2009) Energy Gaps in Etched Graphene Nanoribbons. *Phys. Rev. Lett.* **102**:056403 – 056407.
doi: 10.1103/PhysRevLett.102.056403.

[83] Ci, L., Xu, Z., Wang, L., Gao, W., Ding, F., Kelly, K.F., Yakobson, B.I. and Ajayan, P.M. (2008) Controlled nanocutting of graphene. *Nano Res.* **1**:116 – 122.
doi: 10.1007/s12274-008-8020-9.

[84] Campos, L.C., Manfrinato, V.R., Sanchez-Yamagishi, J.D., Kong, J. and Jarillo-Herrero, P. (2009) Anisotropic Etching and Nanoribbon Formation in Single-Layer. *Nano Lett.* **9**:2600 – 2604.
doi: 10.1021/nl900811r.

[85] Kim, K., Sussman, A. and Zettl, A. (2010) Graphene Nanoribbons Obtained by Electrically Unwrapping Carbon Nanotubes. *ACS Nano* **4**:1362 – 1366.
doi: 10.1021/nn901782g

[86] Zhu, Z., Su, D., Weinberg, G. and Schlogl, R. (2004) Supermolecular Self-Assembly of Graphene Sheets: Formation of Tube-in-Tube Nanostructures. *Nano Lett.* **4**:2255 – 2259.
doi: 10.1021/nl048794t.

[87] Jin, W., Fukushima, T., Niki, M., Kosaka, A., Ishii, N. and Aida, T. (2005) Self-assembled graphitic nanotubes with one-handed helical arrays of a chiral amphiphilic molecular graphene. *Proc. Natl. Acad. Sci. U.S.A.* **102**:10801 – 10806.
doi: 10.1073/pnas.0500852102.

[88] Chen, Q., Chen, T., Pan, G.-B., Yan, H.-J., Song, W.-G., Wan, L.-J., Li, Z.-T., Wang, Z.-H., Shang, B., Yuan, L.-F. and Yang, J.-L. (2008) Structural selection of graphene supramolecular assembly oriented by molecular conformation and alkyl chain. *Proc. Natl. Acad. Sci. U.S.A.* **105**:16849 – 16854.
doi: 10.1073/pnas.0809427105.

[89]   Lee, C., Wei, X., Kysar, J. W. and Hone, J. (2008) Measurement of the Elastic Properties and Intrinsic Strength of Monolayer Graphene. *Science* **321**:385 – 388. doi: 10.1126/science.1157996.

[90]   Bunch, J.S., Verbridge, S.S., Alden, J. S., van der Zande, A.M., Parpia, J.M. Craighead, H.G. and McEuen, P.L. (2008) Impermeable Atomic Membranes from Graphene Sheets. *Nano Lett.* **8**:2458 – 2462. doi: 10.1021/nl801457b.

[91]   Gomez-Navarro, C., Burghard, M. and Kern, K. (2008) Elastic Properties of Chemically Derived Single Graphene Sheets. *Nano Lett.* **8**:2045 – 2049. doi: 10.1021/nl801384y.

[92]   Bets, K.V. and Yakobson, B I. (2009) Spontaneous Twist and Intrinsic Instabilities of Pristine Graphene Nanoribbons. *Nano Res.* **2**:161 – 166.

[93]   Viculis, L.M., Mack, J.J. and Kaner, R.B. (2003) A Chemical Route to Carbon Nanoscrolls. *Science* **299**:1361 – 1361. doi: 10.1126/science.1078842.

[94]   Braga, S.F., Coluci, V.R., Legoas, S.B., Giro, R., Galvao, D.S. and Baughman, R. H. (2004) Structure and Dynamics of Carbon Nanoscrolls. *Nano Lett.* **4**:881 – 884. doi: 10.1021/nl0497272.

[95]   Yu, D. and Liu, F. (2007) Synthesis of Carbon Nanotubes by Rolling up Patterned Graphene Nanoribbons Using Selective Atomic Adsorption. *Nano Lett.* **7**:3046 – 3050. doi: 10.1021/nl071511n.

[96]   Sidorov, A., Mudd, D., Sumanasekera, G., Ouseph, P.J., Jayanthi, C.S. and Wu, S.-Y. (2009) Electrostatic deposition of graphene in a gaseous environment: a deterministic route for synthesizing rolled graphenes? *Nanotechnology* **20**:055611 – 055615. doi: 10.1088/0957-4484/20/5/055611.

[97]   Patra, N., Wang, B. and Král, P. (2009) Nanodroplet Activated and Guided Folding of Graphene Nanostructures. *Nano Lett.* **9**:3766 – 3771. doi: 10.1021/nl9019616.

[98]   Lee, K.-B., Park, S.-J., Mirkin, C.A., Smith, J.C. and Mrksich, M. (2002) Protein Nanoarrays Generated By Dip-Pen Nanolithography. *Science* **295**:1702 – 1705. doi: 10.1126/science.1067172.

[99]   Duwez, A.-S., Cuenot, S., Jerome, C., Gabriel, S., Jerome, R., Rapino, S. and Zerbetto, F. (2006) Mechanochemistry: targeted delivery of single molecules. *Nat. Nanotechnol.* **1**:122 – 125. doi: 10.1038/nnano.2006.92.

[100] Patra, N., Song, Y. and Král, P. (2010) Self-assembly of Graphene Nanostructures on Nanotubes. *ACS Nano*. Online February 22, 2011.

Beilstein-Institut

A Single Molecule Nanoactuator: Toward a Device for Drug Discovery at the Limits of Sensitivity

James Youell[1,#] and Keith Firman[2,*]

[1]IBBS Biophysics Laboratories, School of Biological Sciences, University of Portsmouth, St. Michel's Building, White Swan Road, Portsmouth, PO1 2DT, United Kingdom.

[2]School of Biological Sciences, University of Portsmouth, King Henry Building, King Henry I Street, Portsmouth, PO1 2DY, United Kingdom.

E-Mail: *keith.firman@port.ac.uk or #jim.youell@port.ac.uk

Received: 22nd June 2010 / Published: 13th June 2011

Abstract

This work was initiated by single molecule studies of the molecular motor activity of the Type I Restriction – Modification enzyme *Eco*R124I at a time when the consensus viewpoint was that such motors could not manipulate microscale objects. Type I Restriction-Modification enzymes process DNA, prior to cleavage, by means of translocation and this work was described using such single molecule studies involving the use of a Magnetic Tweezer setup (which also demonstrated that a micron-sized magnetic bead could be pulled over several microns distance by the 20 nm motor).

Recently, this initial work has led to the development of an electronic version of the Magnetic Tweezer setup, which can also manipulate the DNA-attached bead, allowing its use as a biosensor and tool for drug discovery. The system can be used for a wide range of DNA-manipulating enzymes, many of which are potential drug targets.

INTRODUCTION

This work was inspired by studies with a novel molecular motor that is a Type I Restriction-Modification enzyme and it is from this early work that the questions about the capability of such motors first arose. In particular, our studies with *Eco*R124I [1] made it possible to show that biological molecular motors could be used to manipulate materials on the microscale and be used within artificially manufactured devices. However, as will become clear during this paper, our studies have moved away from the R-M enzyme toward other DNA manipulating enzymes as a way to seek commercial development of the device we have produced.

Type I Restriction-Modification Systems

Restriction and Modification (R-M) systems are bacterial protective systems that reduce horizontal transfer of DNA by recognising incoming DNA as "foreign", while "marking" the host DNA as "self" [2]. The most obvious example is illustrated by bacteriophage infection where the bacteriophage DNA is the "foreign", incoming, DNA, which is destroyed by the restriction enzyme (Figure 1). Modification provides the mechanism by which the host DNA is protected from restriction activity. Therefore, restriction enzymes are endonucleases, which recognise and bind at specific DNA sequences and then cleave the DNA unless the protective mechanism of modification prevents this [3]. Modification adds a methyl group to this specific sequence, which prevents cleavage by the associated restriction enzyme, affording the required protection to the host chromosome (Figure 1).

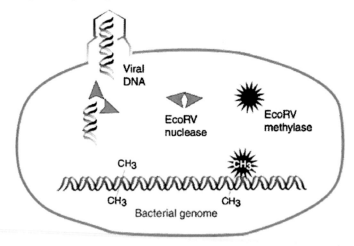

Figure 1. The processes of restriction and modification
The two processes associated with restriction and modification systems involve DNA cleavage by an endonuclease (R.*Eco*RV in this case) and DNA methylation by a DNA methyltransferase (M.*Eco*RV in this case). Methylation occurs within the same DNA binding sequence recognized by the endonuclease and prevents subsequent cleavage at that site.

Type I R-M systems were the first such enzymes identified and unlike the more common Type II enzymes used in Genetic Engineering, they combine the restriction and modification activities into a single, multifunctional, multi-subunit complex [4]. This presents an interesting regulation problem for the enzyme, which must be able to switch between the two activities in a controlled manner [5, 6]. The most obvious control occurs upon DNA-binding and involves the determination of the modification status of the DNA target sequence. If the target is methylated on both strands then the enzyme simply dissociates (this DNA is normally the host chromosomal DNA), while if the DNA is hemi-methylated (methylated on only one strand) then the R-M enzyme acts as a DNA methyltransferase and methylates the unmethylated strand of the target DNA (this would also, normally, be host chromosome DNA, but is hemi-methylated following DNA replication). However, if neither strand of the target DNA is methylated the R-M enzyme undergoes an ATP-dependent conformational change that switches the enzyme to an endonuclease and the DNA is cleaved in an ATP-dependent manner [7]. Cleavage by Type I R-M enzymes is a random process that results in fragments that are generally of random length [8] following a process that involves DNA translocation (Figure 2) [9, 10]. Cleavage appears to occur on a two-site DNA molecule following collision of two translocating enzymes [11], however, on single site circular DNA cleavage is independent of such collisions and appears to follow translocation of "all" of the available DNA [8, 12] suggesting that anything that prevents translocation induces the cleavage reaction [13].

Type I R-M enzymes consist of three different subunits encoded by the following genes [14]:

hsdS: DNA **S**pecificity –   this gene encodes the subunit responsible for DNA sequence recognition and binding.

hsdM: DNA **M**odification – this gene encodes the subunit responsible for DNA methylation using the cofactor *S*-adenosyl methionine (AdoMet or SAM).

hsdR: DNA **R**estriction –   DNA translocation and cleavage.

The subunits produced from these three genes are arranged with a stoichiometry of $HsdR_2:HsdM_2:HsdS_1$ (or $R_2M_2S_1$) in the fully functional holoenzyme (R-M enzyme), however, the subunit assembly pathways are complex and enzyme-dependent with a number of possible sub-assemblies [15, 16]. These sub-assemblies have, in some cases, a role in the control of the restriction and modification activities of the enzyme [5].

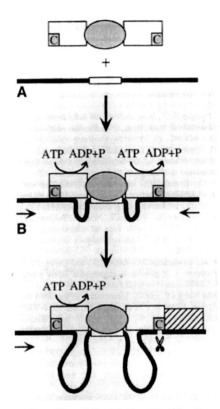

Figure 2. DNA cleavage by a Type I Restriction-Modification enzyme
DNA cleavage by a Type I R-M enzyme involves DNA translocation followed by cleavage at a random site. The core MTase (HsdM$_2$:HsdS$_1$) is illustrated by the grey oval and this complex will bind two HsdR subunits (illustrated by the clear squares) to produce the fully functional R-M enzyme. Binding of this enzyme to an unmethylated recognition sequence (clear rectangle on the DNA, which is represented by a thick black line), in the presence of the cofactor ATP results in DNA translocation (pulling of the DNA) through the bound complex. This continues until a blockage is met (represented by the hatched square) when translocation is stopped and cleavage occurs – this results in a random position of the cleavage site, relative to the binding site, on each molecule as initiation of translocation is asynchronous.

Type I R-M system are divided into four families based on complementation assays, cross-antigenic interactions, biochemical properties, or gene order [3, 17–19] – Type IA [20] (typified by the *Eco*KI and *Eco*BI systems), Type IB [21] (typified by *Eco*AI and *Cfr*A), Type IC [22] (typified by *Eco*R124I and *Eco*DXXI) and Type ID [23] (*Kpn*AI). Type IA, IB and ID are all located on the chromosome, but the first members of the Type IC systems were isolated from conjugative plasmids [24]. However, more recent members of this family have been isolated from the chromosome [25, 26]. The location of the *Eco*R124I and *Eco*R124II R-M systems on the conjugative plasmid R124 present a particular problem for control of restriction versus modification activities as this control must be temporal and modification of the new host chromosome, following conjugal transfer, must occur before any restriction activity [27, 28] and, as mentioned above, for $EcoR124I$ this control appears to involve control of subunit assembly with the final assembly step with the second HsdR subunit being the key stage [5]. The $HsdR_1:HsdM_2:HsdS_1$ sub-assembly complex retains methylation activity and translocation activity [29], but lacks endonuclease activity and the weak P_{RES} promoter limits production of the HsdR subunit, enabling this temporal control.

Magnetic Tweezer Technology

The simplest device for measuring DNA translocation, such as that produced by Type I R-M enzymes, is a Magnetic Tweezer device. Magnetic Tweezers allow real time monitoring of protein-DNA interactions without surface interference and with femtonewton sensitivity. In addition, these systems can measure DNA displacements as low as 10 nm as well as being able to produce negative, or positive, supercoils into the DNA, one turn at a time, through manipulation (spinning) of the magnetic bead [30, 31].

In the study of DNA associated proteins and enzymes the Magnetic Tweezer technique has been found to be a powerful tool [32], it comprises a single DNA molecule anchored at one end to a surface and, at the other end to a micron-scale magnetic bead [33 – 38]. Magnets are used to pull and rotate the microbead, thus stretching and twisting the DNA molecule. The vertical magnetic force causes the DNA to extend and provides a restoring force to restrict the bead's transverse Brownian fluctuations. DNA translocation, winding or twisting induces vertical movement of the bead that can be measured by the Magnetic Tweezer setup.

Comparison of helicases and translocators

The $EcoR124I$ motor was used to demonstrate motor activity at the single molecule level, but more recently our interest has moved to DNA helicases. However, there is a close relationship between Type I R-M enzymes and DNA helicases [39]. Although the docked HsdR subunit of $EcoR124I$ has primarily been described as a translocase in the literature [40], primary sequence analysis suggests that this motor subunit shares a great deal of sequence identity to the Superfamily 2 (SF2) helicases and would be classified under the scheme proposed by Singleton $et\ al.$ [41] as a SF2β helicase (the β denoting the requirement for a double stranded substrate).

The difference between the helicases and translocases are indeed primarily based upon the biochemical characteristics of the system, with both helicases and translocases moving directionally along DNA strands (either single or double stranded) utilising ATP as an energy source, with the main difference being that helicases unwind the substrate and translocases simply track it. The motor domain responsible for ATP hydrolysis is also highly conserved between the two types of motor, with both types utilising two RecA-like core domains for ATP hydrolysis and movement.

Both translocases and helicases often have secondary functions, the *Eco*R124I translocating complex being a Type IC restriction endonuclease has coupled endonuclease activity, whereas the non-structural protein 3 (NS 3) of the hepatitis C virus, possibly the most well researched SF2 helicase couples its' helicase domain with an N-terminal protease domain that is required for processing of viral polyprotein [42].

These similarities shared between the translocases and helicase have allowed us to broaden the reach of our research and therefore the device to allow the study of SF2 helicases from *Plasmodium falciparum* as potential anti-malaria targets.

RESULTS AND DISCUSSION

This paper represents a review and short history of two EC-funded projects (MOL SWITCH) and BIONANO-SWITCH), which were born out of a question set by KF in the late 1990's – *"can we detect the movement produced by a biological molecular motor through its ability to manipulate microscale objects"*. Of course, this concept was driven by the elegant work on myosin [43, 44], where single molecule experiments were used to characterise the mode of action of a motor that can influence movement on the macroscale.

The first step along the path, which we will describe in this paper, was a bid for funding to detect motor activity through induced motion of DNA-attached nanoscale gold particles. However, this first step was quickly halted by a referee's comments that *"Stoke's Law indicated that this was not possible as the motor could not generate sufficient force to manipulate such objects"*. However, despite this major set-back, a chance meeting between KF, Prof. Cees Dekker and Prof. David Bensimon, at a Bionanotechnology Conference in Berkley, California, led to the idea of measuring motor activity for a Type I R-M enzyme using a Magnetic Tweezer setup and this work built upon the knowledge generated by bulk-solution, biochemical studies.

Measurement of DNA Translocation

➡ 5'-**GAA**TTCGAG**GTCG**-3' *EcoR124I recognition sequence*

⇒ 5'-YAAGAAAGAAAGAAGAAAGAAAY-3' *Triple-helix binding oligonucleotide*

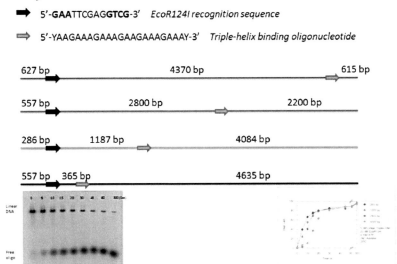

Figure 3. The concept of the triple-helix displacement assay
Four plasmids were prepared with binding sites for the *Eco*R124I enzyme toward one end (black arrows) and triple-helix binding sites at variable distances from the *Eco*R124I binding site (green arrows). Translocation by the *Eco*R124I motor resulted in displacement of a radiolabelled triple-forming oligonucleotide (agarose gel inset at left) and the concentration of bound triplex-forming oligonucleotide against time produced the inset graph showing lag periods at the beginning of displacement.

The first measurements of DNA translocation by the Type I R-M enzyme *Eco*R124I were a collaborative effort with Dr. Mark Szczelkun at the University of Bristol. The work was based on an assay using a triple helix displacement assay developed by Szczelkun [45] in which the distance between the DNA-binding sequence of the *Eco*R124I enzyme and a triple helix binding site was varied (Figure 3). Displacement of the triple-helix-binding oligonucleotide was measured with a radiolabelled sample and the results analysed using agarose gel electrophoresis (Figure 3 inset). Analysis of the data, using a multistep model for translocation and measuring the lag periods for displacement, determined that the speed of translocation was ~400 bp s^{-1}. This provided much of the required background knowledge that was required for single-molecule analysis of translocation and allowed a comparison of the kinetics measured using single molecules as against bulk-solution measurements.

The first step toward single molecule studies with *Eco*R124I was the use of an Atomic Force Microscope (AFM) to analyse the motor bound to DNA and the quality of some of the data obtained from Dekker's laboratory is illustrated (Figure 4), which encouraged us to measure loop-sizes for the translocating molecules and determine the rate of translocation [46]. However, the results obtained were some 10-fold lower than expected due to errors in time measurement with this process. Despite this drawback, the AFM analysis also allowed us to closely examine the initiation complex for translocation (obtained by using the non-hydro-lysable analogue of ATP – ATPγS) and identify an ssDNA "bubble" associated with initiation of translocation [46].

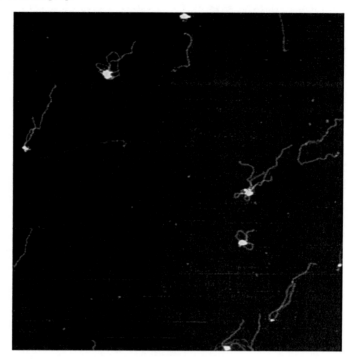

Figure 4. AFM analysis of *Eco*R124I bound to DNA
Images of dsDNA with bound *Eco*R124I enzyme showing loops that are indicative of DNA translocation. There are also some mini-aggregates where several *Eco*R124I enzymes are bound together and a few examples of only the bound MTase.

The AFM work showed that the protein was very clean, bound DNA as expected and seemed to translocate DNA as expected (we believed that the slow translocation rate was an artefact of the timing mechanism following fixing of the complexes onto a mica surface). Therefore, we initiated work with the Magnetic Tweezer setup that could measure the speed of translocation and the forces generated [47]. A typical trace from the Magnetic Tweezer experiments is shown in Figure 5 and shows a motor that has high processivity, moves rapidly from stationary to full speed and resets after a random time period [47]. Analysis of the slope of these traces gave a speed for a single-motor subunit (HsdR) of 450 bp s^{-1}, which is very close to the figure obtained in the bulk experiments (where many of the motors will be idle), which was a very re-assuring situation. Force measurement provided unexpected results as increasing the force on the DNA (pulling opposite to the motor activity) did not stall the motor, but rather decreased processivity until at about 8 pN applied force there was little or no translocation observed. This force is very similar to that generated by motors such as myosin [48] and kinesin [49]. Further, a more detailed analysis using the Magnetic

Tweezer setup allowed us to demonstrate how the stalling of the motor led to dissociation of the motor unit and that the two motor subunits, present in the *Eco*R124I holoenzyme [16], were independently functional motors [29].

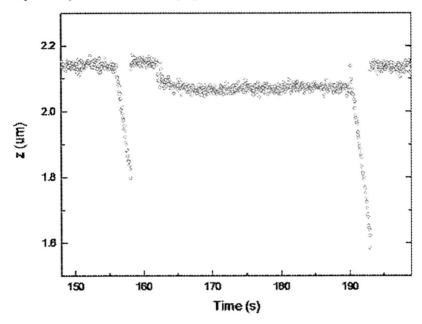

Figure 5. Output from a Magnetic Tweezer analysis of DNA translocation.
A typical trace from a Magnetic Tweezer setup where a 4 kbp linear DNA molecule was attached to the surface of the flow cell through a PCR product, containing 10% Digoxygenin-dUTP, ligated to the plasmid and immobilised on a monolayer of anti-DIG antibodies. At the opposite end of the plasmid is another PCR product, incorporating biotinylated nucleotides, also ligated to the plasmid and the attached to a streptavidin-coated Dynal™ 3 micron bead.

In parallel with this work, David Bensimon's group were also investigating the motor activity of another enzyme that could translocate DNA – FtsK – also using a Magnetic Tweezer setup and they were able to show that FtsK is the Ferrari of the two motors with a speed that is 10-fold that of *Eco*R124I (approaching $2\,\mu m\ s^{-1}$). In addition, FtsK does not require a specific binding site, but attaches to almost any DNA and then pulls the DNA through the bound complex in a highly processive manner [50].

This work (funded through the EC grant MOL SWITCH – http://www.nanonet.org.uk/molswitch) had now answered the question set by KF – the two DNA translocating enzymes were motors that could indeed manipulate micron-scale objects (the beads used in the Magnetic Tweezer setup varied between 1 and $3\,\mu m$ diameter and, perhaps even more interesting, the longest "run" observed moved these beads $> 3\,\mu m$ distance – governed by the height of the flowcell, or the total length of the DNA substrate). Therefore, the scene was set for the next question: *Could a biological molecular motor be used as a nanoactuator within a useful device?*

The use of *Eco*R124I as one potential motor within a nanoactuator presented an interesting problem – DNA cleavage – the normal activity of *Eco*R124I, to cleave unmethylated DNA was a potential problem within a magnetic Tweezer setup as this cleavage would result in loss of beads, but our initial experiments were performed with the previously described subunit-assembly intermediate complex of *Eco*R124I, which has only one HsdR subunit per MTase – an R_1-complex ($R_1M_2S_1$) that does not cleave DNA. Interestingly, although stalling of the enzyme, even as the complete holoenzyme (R_2-complex), was frequently seen in the Magnetic Tweezer, DNA cleavage was a rare event. However, to overcome any such problems, as part of the MOL SWITCH Project, cleavage deficient mutants of HsdR were prepared by one of the Partner Groups [51]. Therefore, we now had two motors that were able to act as nanoactuators within any proposed device and had shown that we could detect movement of a bead using an optical Magnetic Tweezer Setup.

An electronic Magnetic Tweezer

The first step toward using biological molecular motors within a biosensing device was to simplify the optical Magnetic Tweezer setup, which was primarily designed as a tool for research laboratories (see http://www.picotwist.com) and the concept (Figure 6A) we imagined was a biological molecular dynamo, where a biological motor would move a magnetic bead within range of a sensor that could detect this movement in the vertical plane. It was clear that two types of sensor were possible – a magnetoresistive device [52] or a Hall Effect device [53] and two new partners, working with such sensors, were added to the consortium involved in the MOL SWITCH Project to develop a new proposal.

At about the same time there was an EC call for projects in the area of Synthetic Biology and it was clear that developing a generic biosensor that uses a biological molecular motor as the transducer of the biosensor, outputting an electronic signal from detection of a biological event, would be Synthetic Biology. Such a device would require construction of novel biomaterials that would allow surface attachment of biomolecules and would be a clear example of integration of silicon-based devices with single biomolecules. From this concept the EC-funded project BIONANO-SWITCH was born.

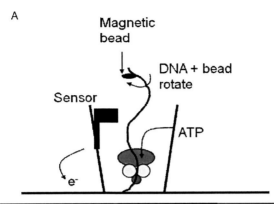

A

Magnetic bead

DNA + bead rotate

Sensor

ATP

e⁻

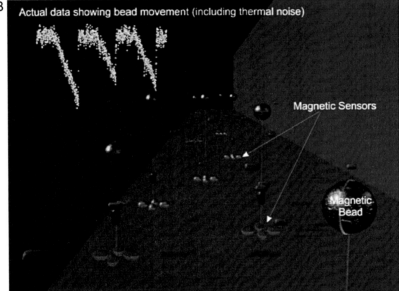

B   Actual data showing bead movement (including thermal noise)

Magnetic Sensors

Magnetic Bead

Figure 6. The "MOL SWITCH" Device
(A) A simple cartoon that was used to suggest the concept of the original MOL SWITCH Project. In the proposed device a DNA molecule is attached to a surface, at one end, and to a magnetic bead at the opposite end in a manner analogous to that used in a Magnetic Tweezer setup. The motor can be introduced, will bind and translocate the DNA, pulling the magnetic bead past an electronic sensor, which outputs an electrical signal, acting as a molecular dynamo. **(B)** This image is taken from an animation available at www.bionano-switch.info and shows a view inside of a flowcell of a Lab-on-a-Chip device that incorporates the proposed Electronic Magnetic Tweezer device. Sensors are located on the floor of the flowcell and attached above these sensors is a single DNA molecule attached to a Magnetic bead. The DNA is stretched using an external magnetic field and movement of the bead, by a molecular motor, is detected by the sensor (indicated as a red sensor).

A Synthetic Biology approach to a biosensor design

In order to construct a generic biosensor, which would employ some of the concepts of Synthetic Biology during development, we proposed a modular construction that "bolted" a single-molecule nanoactuator as the transducer (output) onto an existing detection technology – the Sandwich ELISA – using a high throughput system [54]. The proposed device employed a "Molecular Amplifier" to both improve the signal-noise ratio of the ELISA and to enable release of a biological motor, from an immobilised form, following antigen detection within the ELISA setup. In addition, we imagined the use of DNA aptamers [55] as replacement for the antibodies within the ELISA system which would make the biosensor totally generic (in theory aptamers could be raised to almost any ligand).

The bulk of the Genetic Engineering aspects of the BIONANO-SWITCH Project involved construction of a "cassette like" or "Lego" system, to mimic systems previously described in this area [56], to link the proteins of interest to other moieties (e. g. biotin, Glutathione-S-Transferase and proteases). Eventually all of the biomaterials were created, although with a few problems of protein insolubility for specific constructions – an important aspect of the "Lego" approach for Synthetic Biology is that biomaterials cannot be "engineered" in a way that is more feasible with chemicals and certain protein-fusion constructs can behave in unpredictable ways – trials of the Generic Biosensor/Molecular Amplifier concepts demonstrated that while the system worked, there was no amplification of the signal. The driving force behind this work was to demonstrate the application of Synthetic Biology techniques within a commercially viable device, but the complexity of the system and the unexpected low turnover of one of the fusion proteins make commercial exploitation of this seem unlikely. This is also an important message for the future of devices based around Synthetic Biology – the unpredictable behaviour of novel bio-constructs suggests that an important aspect of the iGem [56] "Lego-like" approach to the production of novel proteins requires a background database of observations with such constructs, which must include unexpected results including reduced solubility and reduced function of protein fusions. Without such an information store many devices will be labour intensive during the development stages, may repeat existing data or knowledge and result in failure.

The proposed nanoactuator device

Despite the above comments, one aspect of the BIONANO-SWITCH Project was very successful, which involved integration of biomaterials, including the biological molecular motors, within a silicon-based sensor. Perhaps this is one aspect of Synthetic Biology that will be commercially viable and this reflects both robustness and reliability of biological molecular machines.

The nanoactuator device will be described in detail elsewhere, but the concept of the device can be understood from Figure 6B. The transducer element is a single biomolecular motor that manipulates a single magnetic bead attached to a single DNA molecule. In effect the sensor is a single molecule switch, which provides an on-off signal of motor activity. Once again our interest in such a device was driven by the concept of commercial application of the device and market research carried out in Portsmouth indicated the most lucrative market might be drug discovery. Many biological molecular motors (topoisomerases, helicases, polymerases etc.) are drug targets and the advantages that can be gained by using single molecule studies have already been clearly demonstrated with topoisomerases [57, 58]. In the proposed device we can readily imagine a multiple channel system (illustrated with a two channel setup in Figure 7A) where one channel can detect normal motor activity, while the other channels can be used to detect drug-induced loss of activity (Figure 7B). Our own work has focussed on the use of helicases from *Plasmodium falciparum* that have been proposed as potential antimalarial drug targets [59], but this work is currently at a very early stage.

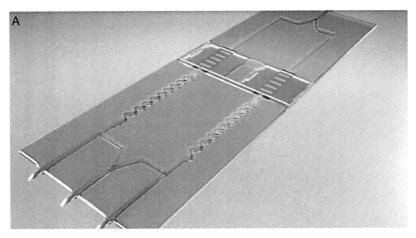

Figure 7. The proposed nanoactuator device.
(A) A two-channel flowcell of the proposed commercial version of the Electronic Magnetic Tweezer in which one can see the attachment sites for the DNA molecules (gold pads) that are fed buffer, motor and drugs through the microfluidics.

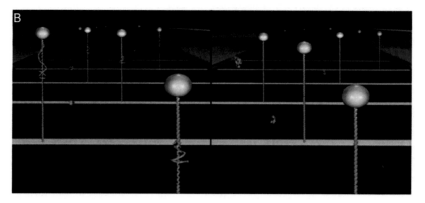

Figure 7. The proposed nanoactuator device.
(B) A view of the above flowcell, side by side, showing DNA+beads attached to the gold pads and functioning helicase molecules in the left channel unwinding DNA from a gapped substrate, but helicase+drug molecules in the right channel that show loss of function. Electronic signals from the device measure loss of activity from the drug-target interaction at the single molecule level and, therefore, at a sensitivity based on the equilibrium constant of the drug-target interaction.

CONCLUSION

Biological molecular motors are surprisingly robust and very powerful. The (early) concept that such motors could not manipulate objects on the microscale was both wrong and surprising as a viewpoint, but several groups have now been able to disprove this assumption and this suggests a potential for such motors in useful devices. This idea is one of the more promising aspects of Synthetic Biology, where integration of biomaterial within silicon devices has been shown to be surprisingly successful and we have also been able to demonstrate efficient surface attachment of biomolecules, at precise location (< 100 nm error), at the single molecule level. Our attempts to apply the "cassette" approach for the construction of novel protein fusions was very mixed and protein insolubility, or loss of function, was a major issue, but was also unpredictable – we suggest that the eventual application of such techniques within Synthetic Biology requires a detailed database of results to avoid repetitive failures for specific constructs. Finally, we believe we have developed a single molecule biosensor, which can be used in the area of drug discovery, that detects drug-target interactions at the single molecule level where the sensitivity of the system is determined by the chemical equilibrium constant of the interaction (actually by k_{off}). We believe this system offers an opportunity to screen both existing and new drugs at a level not previously possible against DNA manipulating targets. Adaptation of the device to novel targets (e. g. protein-protein interactions and chromatin remodelling) provides a huge potential for future application of the technology.

ACKNOWLEDGEMENTS

This work was supported by EC funding through the FET-OPEN scheme (MOL SWITCH, Grant No IST-2001–38036) and the NEST Pathfinder (Synthetic Biology) scheme (BIO-NANO-SWITCH Grant No 043288). We would like to thank all of our collaborators from both of these Projects for their input to this work. Finally, KF would like to thank Martin Hicks for the opportunity to present this work at the recent Beilstein Symposium on Functional Nanoscience.

REFERENCES

[1]   Youell, J. & Firman, K. (2008) *Eco*R124I: from plasmid-encoded restriction modification system to nanodevice. *Microbiol. Mol. Biol. Rev.* **72**:365–377.
       doi: 10.1128/MMBR.00043-07.

[2]   Price, C. & Bickle, T.A. (1986) A possible role for restriction in bacterial evolution. *Micro. Sci.* **3**:296–299.

[3]   Wilson, G.G. & Murray, N.E. (1991) Restriction and modification systems. *Ann. Rev. Gen.* **25**:585–627.
       doi: 10.1146/annurev.ge.25.120191.003101.

[4]   Eskin, B. & Linn, S. (1972) The deoxyribonucleic acid modification and restriction enzymes of *Escherichia coli* B. II. Purification subunit structure and catalytic properties of the restriction endonuclease. *J. Biol. Chem.* **247**:6183–6191.

[5]   Firman, K., Dutta, C., Weiserova, M. & Janscak, P. (2000) The role of subunit assembly in the functional control of type I restriction-modification enzymes. *Mol. Biol. Today* **1**:1–8.

[6]   Burckhardt, J., Weisemann, J. & Yuan, R. (1981) Characterisation of the DNA methylase activity of the restriction enzyme from *Escherichia coli* K. *J. Biol. Chem.* **256**:4024–4032.

[7]   Yuan, R., Bickle, T.A., Ebbers, W. & Brack, C. (1975) Multiple steps in DNA recognition by restriction endonuclease from *E. coli* K. *Nature* **256**:556–560.
       doi: 10.1038/256556a0.

[8]   Szczelkun, M.D., Janscák, P., Firman, K. & Halford, S.E. (1997) Selection of non-specific DNA cleavage sites by the type IC restriction endonuclease *Eco*R124I. *J. Mol. Biol.* **271**:112–123.
       doi: 10.1006/jmbi.1997.1172.

[9]   Yuan, R., Hamilton, D.L. & Burckhardt, J. (1980) DNA translocation by the restriction enzyme from *E. coli* K. *Cell* **20**:237–244.

[10]   Yuan, R., Heywood, J. & Meselson, M. (1972) ATP hydrolysis by restriction endonuclease from *E. coli* K. *Nat. New Biol.* **240**:42–43.

[11]   Studier, F.W. & Bandyopadhyay, P.K. (1988) Model for how type I restriction enzymes select cleavage sites in DNA. *Proc. Natl. Acad. Sci. U.S.A.* **85**:4677–4681. doi: 10.1073/pnas.85.13.4677.

[12]   Szczelkun, M.D., Dillingham, M.S., Janscák, P., Firman, K. & Halford, S.E. (1996) Repercussions of DNA tracking by the type IC restriction endonuclease *Eco*R124I on linear, circular and catenated substrates. *EMBO J.* **15**:6335–6347.

[13]   Janscak, P., MacWilliams, M.P., Sandmeier, U., Nagaraja, V. & Bickle, T.A. (1999) DNA translocation blockage, a general mechanism of cleavage site selection by type I restriction enzymes. *EMBO J.* **18**:2638–2647. doi: 10.1093/emboj/18.9.2638.

[14]   Glover, S.W. & Colson, C. (1969) Genetics of host-controlled restriction and modification in *Escherichia coli*. *Gen. Res. (Cambridge)* **13**:227–240. doi: 10.1017/S0016672300002901.

[15]   Dryden, D.T.F., Cooper, L.P., Thorpe, P.H. & Byron, O. (1997) The *in vitro* assembly of the *Eco*KI type I DNA restriction/modification enzyme and its *in vivo* implications. *Biochemistry* **36**:1065–1076. doi: 10.1021/bi9619435.

[16]   Janscák, P., Dryden, D. & Firman, K. (1998) Analysis of the subunit assembly of the type IC restriction-modification enzyme *Eco*R124I. *Nucl. Acids Res.* **26**:4439–4445. doi: 10.1093/nar/26.19.4439.

[17]   Wilson, G.G. (1991) Organisation of restriction-modification systems. *Nucl. Acids Res.* **19**:2539–2566. doi: 10.1093/nar/19.10.2539.

[18]   Daniel, A.S., Fuller-Pace, F.V., Legge, D.M. & Murray, N.E. (1988) Distribution and diversity of *hsd* genes in *Escherichia coli* and other enteric bacteria. *J. Bacteriol.* **170**:1775–1782.

[19]   Murray, N.E., Gough, J.A., Suri, B. & Bickle, T.A. (1982) Structural homologies among type I restriction and modification systems. *EMBO J.* **1**:535–539.

[20]   Barcus, V.A., Titheradge, A.J.B. & Murray, N.E. (1995) The diversity of alleles at the *hsd* locus in natural populations of *Escherichia coli*. *Genetics* **140**:1187–1197.

[21] Fuller-Pace, F.V., Cowan, G.M. & Murray, N.E. (1985) *Eco*A and *Eco*E: alternatives to the *Eco*K family of type I restriction and modification systems of *Escherichia coli*. *J. Mol. Biol.* **186**:65 – 75.
doi: 10.1016/0022-2836(85)90257-8.

[22] Price, C., Pripfl, T. & Bickle, T.A. (1987) *Eco*R124 and *Eco*R124/3: the first members of a new family of type I restriction and modification systems. *Eur. J. Biochem.* **167**:111 – 115.
doi: 10.1111/j.1432-1033.1987.tb13310.x.

[23] Lee, N.S., Rutebuka, O., Arakawa, T., Bickle, T.A. & Ryu, J. (1997) *Kpn*AI, a new type I restriction-modification system in *Klebsiella pneumoniae. J. Mol. Biol.* **271**:342 – 8.
doi: 10.1006/jmbi.1997.1202.

[24] Firman, K. & University of Newcastle upon Tyne. (1982) *The restriction and modification systems coded by the bacterial plasmid R124*, University of Newcastle upon Tyne, UK.

[25] Piekarowicz, A., Skrzypek, E. & Goguen, J.D. (1985) The *Eco*DXXI restriction and modification system of *Escherichia coli* ET7. In *Gene manipulation and expression* (Glass, R.E. & Spizek, J., eds.), pp. 79 – 94. Croom Helm, London.

[26] Tyndall, C., Meister, J. & Bickle, T.A. (1994) The *Escherichia coli prr* region encodes a functional type IC DNA restriction system closely integrated with an anticodon nuclease gene. *J. Mol. Biol.* **237**:266 – 274.
doi: 10.1006/jmbi.1994.1230.

[27] Prakash-Cheng, A., Chung, S.S. & Ryu, J. (1993) The expression and regulation of hsd_K genes after conjugative transfer. *Mol. Gen. Gen.* **241**:491 – 496.
doi: 10.1007/BF00279890.

[28] Prakash-Cheng, A. & Ryu, J. (1993) Delayed expression of *in vivo* restriction activity following conjugal transfer of *Escherichia coli* hsd_K (restriction-modification) genes. *J. Bacteriol.* **175**:4905 – 4906.

[29] Seidel, R., Bloom, J.G.P., van Noort, J., Dutta, C.F., Dekker, N.H., Firman, K., Szczelkun, M.D. & Dekker, C. (2005) Dynamics of initiation, termination and reinitiation of DNA translocation by the motor protein *Eco*R124I. *EMBO J.* **24**:4188 – 4197.
doi: 10.1038/sj.emboj.7600881.

[30] Strick, T., Allemand, J.F., Bensimon, D., Lavery, R. & Croquette, V. (1999) Twisting and Stretching a DNA Molecule Leads to Structural Transitions. *AIP Conference Proceedings*, 249 – 270.
doi: 10.1063/1.59890.

[31]   Strick, T.R., Croquette, V. & Bensimon, D. (2000) Single-molecule analysis of DNA uncoiling by a type II topoisomerase. *Nature* **404**:901 – 904. doi: 10.1038/35009144.

[32]   Kim, K. & Saleh, O.A. (2009) A high-resolution magnetic tweezer for single-molecule measurements. *Nucl. Acids Res.* **37**:e136-.

[33]   Chiou, C.-H., Huang, Y.-Y., Chiang, M.-H., Lee, H.-H. & Lee, G.-B. (2006) New magnetic tweezers for investigation of the mechanical properties of single DNA molecules. *Nanotechnology* 1217 – 1224. doi: 10.1088/0957-4484/17/5/009

[34]   Gosse, C. & Croquette, V. (2002) Magnetic Tweezers: Micromanipulation and Force Measurement at the Molecular Level. *Biophys. J.* **82**:3314 – 3329. doi: 10.1016/S0006-3495(02)75672-5.

[35]   Gupta, P., Zlatanova, J. & Tomschik, M. (2009) Nucleosome Assembly Depends on the Torsion in the DNA Molecule: A Magnetic Tweezers Study. *Biophys. J.* **97**:3150 – 3157. doi: 10.1016/j.bpj.2009.09.032.

[36]   Hosu, B.G., Jakab, K., Banki, P., Toth, F.I. & Forgacs, G. (2003) Magnetic tweezers for intracellular applications. *Rev. Sci. Instrum.* **74**:4158 – 4163. doi: 10.1063/1.1599066.

[37]   Lipfert, J., Hao, X. & Dekker, N.H. (2009) Quantitative Modeling and Optimization of Magnetic Tweezers. *Biophys. J.* **96**:5040 – 5049. doi: 10.1016/j.bpj.2009.03.055.

[38]   Zlatanova, J. & Leuba, S.H. (2003) Magnetic tweezers: a sensitive tool to study DNA and chromatin at the single-molecule level. *Biochem. Cell Biol.* **81**:151 – 159. doi: 10.1139/o03-048.

[39]   Durr, H., Flaus, A., Owen-Hughes, T. & Hopfner, K.-P. (2006) Snf2 family ATPases and DExx box helicases: differences and unifying concepts from high-resolution crystal structures. *Nucl. Acids Res.* **34**:4160 – 4167. doi: 10.1093/nar/gkl540.

[40]   Stanley, L.K., Seidel, R., van der Scheer, C., Dekker, N.H., Szczelkun, M.D. & Dekker, C. (2006) When a helicase is not a helicase: dsDNA tracking by the motor protein *Eco*R124I. *EMBO J.* **25**:2230 – 9. doi: 10.1038/sj.emboj.7601104.

[41]   Singleton, M.R., Dillingham, M.S. & Wigley, D.B. (2007) Structure and mechanism of helicases and nucleic acid translocases. *Annu. Rev. Biochem.* **76**:23 – 50. doi: 10.1146/annurev.biochem.76.052305.115300.

[42]   Mackintosh, S.G., Lu, J.Z., Jordan, J.B., Harrison, M.K., Sikora, B., Sharma, S.D., Cameron, C.E., Raney, K.D. & Sakon, J. (2006) Structural and biological identification of residues on the surface of NS 3 helicase required for optimal replication of the hepatitis C virus. *J. Biol. Chem.* **281**:3528 – 35.
        doi: 10.1074/jbc.M512100200.

[43]   Bagshaw, C.R. & Conibear, P.B. (1998). *Enzyme catalysis: structure, dynamics and chemistry, Leicester.*

[44]   Conibear, P.B., Kuhlman, P.A. & Bagshaw, C.R. (1998) Measurement of ATPase activities of myosin at the level of tracks and single molecules. *Adv. Exp. Med. Biol.* **453**:15 – 26; discussion 26 – 7.

[45]   Firman, K. & Szczelkun, M. (2000) Measuring motion on DNA by the type I restriction endonuclease *Eco*R124I using triplex dissociation. *EMBO J.* **19**:2094 – 2102.
        doi: 10.1093/emboj/19.9.2094.

[46]   van Noort, J., van der Heijden, T., Dutta, C.F., Firman, K. & Dekker, C. (2004) Initiation of Translocation by Type I Restriction-Modification Enzymes is Associated with a Short DNA Extrusion. *Nucl. Acids Res.* **32**:6540 – 6547.
        doi: 10.1093/nar/gkh999.

[47]   Seidel, R., van Noort, J., van der Scheer, C., Bloom, J.G.P., Dekker, N.H., Dutta, C. F., Blundell, A., Robinson, T., Firman, K. & Dekker, C. (2004) Real-Time Observation of DNA Translocation by the Type I Restriction-Modification Enzyme *Eco*R124I. *Nat. Struct. Mol. Biol.* **11**:838 – 843.
        doi: 10.1038/nsmb816.

[48]   Finer, J.T., Simmons, R.M. & Spudich, J.A. (1994) Single myosin molecule mechanics: piconewton forces and nanometre steps. *Nature* **368**:113 – 119.
        doi: 10.1038/368113a0.

[49]   Bormuth, V., Varga, V., Howard, J. & Schaffer, E. (2009) Protein Friction Limits Diffusive and Directed Movements of Kinesin Motors on Microtubules. *Science* **325**:870 – 873.
        doi: 10.1126/science.1174923.

[50]   Saleh, O.A., Perals, C., Barre, F.X. & Allemand, J.F. (2004) Fast, DNA-sequence independent translocation by FtsK in a single-molecule experiment. *EMBO J.* **23**:2430 – 9.
        doi: 10.1038/sj.emboj.7600242.

[51] Sisakova, E., Stanley, L.K., Weiserova, M. & Szczelkun, M.D. (2008) A RecB-family nuclease motif in the Type I restriction endonuclease *Eco*R124I. *Nucl. Acids Res.* **36**:3939–3949.
doi: 10.1093/nar/gkn333.

[52] Graham, D.L., Ferreira, H.A. & Freitas, P.P. (2004). Magnetoresistive-based biosensors and biochips. *TiBtech* **22**:455–462.
doi: 10.1016/j.tibtech.2004.06.006.

[53] Boero, G., Utke, I., Bret, T., Quack, N., Todorova, M., Mouaziz, S., Kejik, P., Brugger, J., Popovic, R.S. & Hoffmann, P. (2005) Submicrometer Hall devices fabricated by focused electron-beam-induced deposition. *Appl. Phys. Lett.* **86**:042503 (online).
doi: 10.1063/1.1856134.

[54] Xu, Z.-W., Zhang, T., Song, C.-J., Li, Q., Zhuang, R., Yang, K., Yang, A.-G. & Jin, B.-Q. (2008) Application of sandwich ELISA for detecting tag fusion proteins in high throughput. *Appl. Phys. Lett.* **81**:183–189.

[55] Blank, M. & Blind, M. (2005) Aptamers as tools for target validation. *Curr. Opin. Chem. Biol.* **9**:336–342.
doi: 10.1016/j.cbpa.2005.06.011.

[56] Heinemann, M. & Panke, S. (2006) Synthetic biology–putting engineering into biology. *Bioinformatics* **22**:2790–9.
doi: 10.1093/bioinformatics/btl469.

[57] Koster, D.A., Palle, K., Bot, E.S.M., Bjornsti, M.-A. & Dekker, N.H. (2007) Antitumour drugs impede DNA uncoiling by topoisomerase I. *Nature* **448**:213–217.
doi: 10.1038/nature05938.

[58] Strick, T.R., Charvin, G., Dekker, N.H., Allemand, J.F., Bensimon, D. & Croquette, V. (2002) Tracking enzymatic steps of DNA topoisomerases using single-molecule micromanipulation. *Comptes Rendus Physique* **3**:595–618.
doi: 10.1016/S1631-0705(02)01347-6.

[59] Tuteja, R. (2007) Helicases; feasible antimalarial drug target for *Plasmodium falciparum*. *FEBS J.* **274**:4699–4704.
doi: 10.1111/j.1742-4658.2007.06000.x.

STRUCTURALLY PERSISTENT MICELLES: THEORY AND EXPERIMENT

CHRISTOF M. JÄGER[1], ANDREAS HIRSCH[2,3], CHRISTOPH BÖTTCHER[4] AND TIMOTHY CLARK[1,3,*]

[1]Computer-Chemie-Centrum and Interdisciplinary Center for Molecular Materials, Department of Chemistry and Pharmacy, University of Erlangen-Nürnberg, Nägelsbachstraße 25, 91052 Erlangen, Germany.

[2]Interdisciplinary Center for Molecular Materials, Department of Chemistry and Pharmacy, University of Erlangen-Nürnberg, Henkestraße 25, 91054 Erlangen, Germany.

[3]Excellence Cluster "Engineering of Advanced Materials", University of Erlangen-Nürnberg, Nägelsbachstraße 49b, 91052 Erlangen, Germany.

[4]Research Center of Electron Microscopy, Institute of Chemistry and Biochemistry, Free University Berlin, Fabeckstraße 36a, 14195 Berlin, Germany.

E-MAIL: *Tim.Clark@chemie.uni-erlangen.de

Received: 20th July 2010 / Published: 13th June 2011

ABSTRACT

We describe the progress made in understanding the factors that determine the size, structure and stability of structurally persistent micelles using a combination of designed synthesis, cryo-TEM imaging and molecular-dynamics simulations. The importance of specific counterion effects is revealed in detail. An unexpected effect of sodium counterions leads to attraction between the polycarboxylate head groups of the tailored dendrimers that make up the micelles. This effect even leads to the formation of "superlattices" of highly negatively charged micelles.

INTRODUCTION

Soft nanostructures can be considered to be the second class of nanomaterials after to "hard" nanoparticles and similar structurally defined and static nanoscale structures. Nanostructured soft matter represents a challenging research area, both for theory and experiment. This is because soft matter is inherently dynamic in its structure and cannot, therefore be treated as a single static object. Nonetheless, soft nanostructures can have significant advantages over hard nanoparticles. They are, for instance, formed in a dynamic equilibrium process, so that their self-assembly is governed by thermodynamics, rather than the less predictable process of kinetically controlled nucleation and precipitation. This advantage can, however, soon become a disadvantage because many factors may determine the delicate equilibrium that gives rise to soft nanostructures, so that they may be sensitive to their environment. Nature uses soft nanoparticles almost exclusively in living organisms, so that there can be no doubt that technological applications based on soft nanostructures are potentially extremely powerful. Soft nanostructures are most likely to self-assemble from organic precursor molecules in solution, probably aqueous. Conventional micelles [1] are perhaps the best known soft nanoparticles, but although a very large amount of data about, for instance critical micelle concentrations is available [2], relatively little is known about the detailed structure and dynamics of micelles and other soft nanostructures.

Structurally persistent micelles [3] therefore represent an important milestone in the science of soft nanostructures as they have consistent, well defined and persistent structures that can be observed in detail by techniques such as cryo-transmission electron microscopy (cryo-TEM). These characteristics not only make structurally persistent micelles intriguing experimental objects, but also make them ideal for testing and validating molecular-dynamics (MD) simulations and the force fields used for them. As we will describe below, simulations have played a major role in advancing our understanding of structurally persistent micelles and the factors that control their stability and structure.

MOLECULAR COMPONENTS

In 2004 Kellermann *et al.* [3] described the aggregation of seven amphiphilic dendrocalixarene molecules **1** to uniform and structurally persistent micelles (Scheme 1).

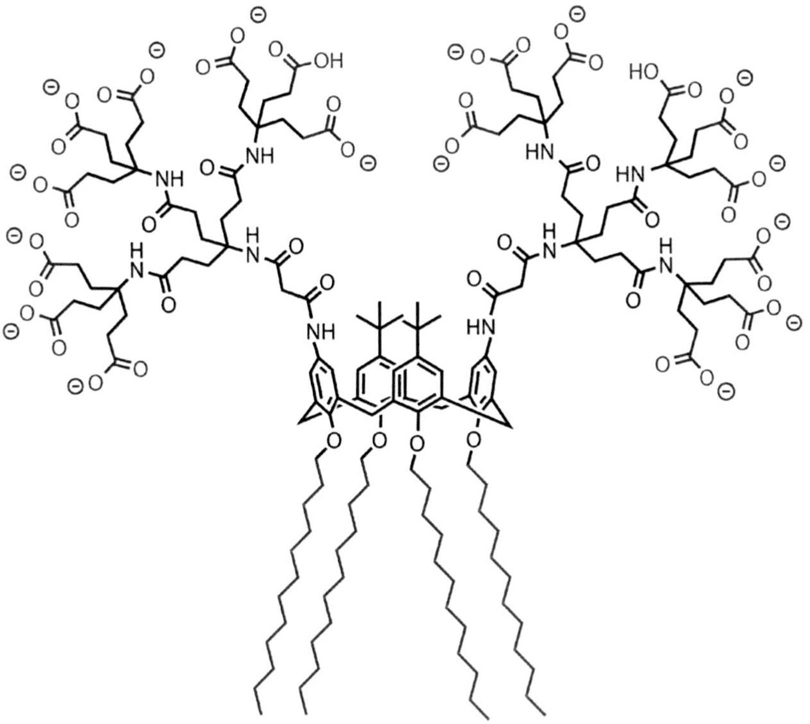

1

Scheme 1. Dendrocalixarene monomer.

The self aggregation behaviour of the molecules was investigated by pulse-gradient spin-echo (PGSE) NMR spectroscopy and cryo-TEM experiments. Remarkably, the experiments showed that a single distinct type of micelle was formed with no other aggregates of different sizes. The key feature of this first amphiphilic dendrimer investigated is its cone-shaped structure, which makes it possible to form small aggregates with high curvature. Each hydrophilic polycarboxylate head-group is linked to four hydrophobic alkane chains by a calyx[4]arene unit.

Hirsch et al. [4 – 7] later synthesized, characterized and investigated the aggregation behaviour of a series of new amphiphilic building blocks based on either calixarenes, fullerenes (Scheme 2: **2, 3**) or later perylene [8, 9] as the central scaffold. These scaffolds allow a variety of hydrophilic and hydrophobic head and tail groups to be bound in a stereochemically controlled and tunable fashion. The polar head groups are Newkome-type oligocarboxylic acids in all cases. At neutral pH, they are predominantly deprotonated and guarantee excellent water solubility.

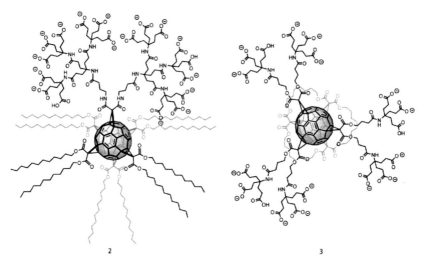

Scheme 2. Fullerene based amphiphilic building blocks.

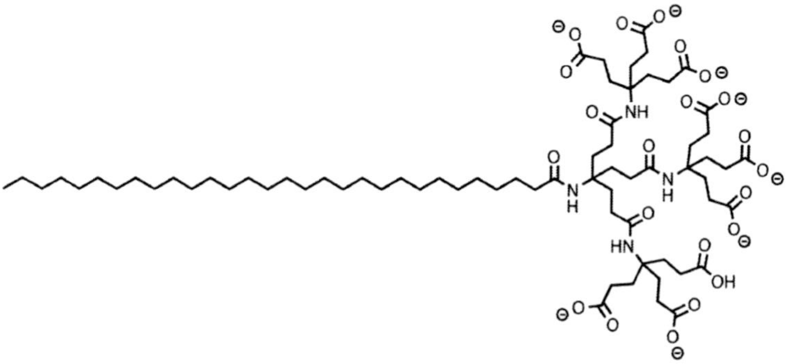

4

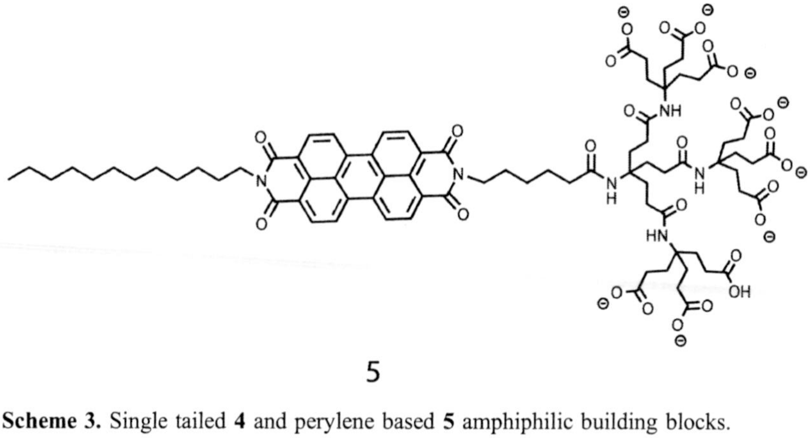

5

Scheme 3. Single tailed **4** and perylene based **5** amphiphilic building blocks.

STRUCTURALLY PERSISTENT MICELLES – CRYO-TEM

As outlined, structurally persistent micelles were first observed [3] by a combination of spectroscopic and microscopic techniques, but the most striking results arise from cryo-TEM studies, which reveal thousands of essentially identical micellar structures, as shown in Figure 1a for the dendrimer **1**.

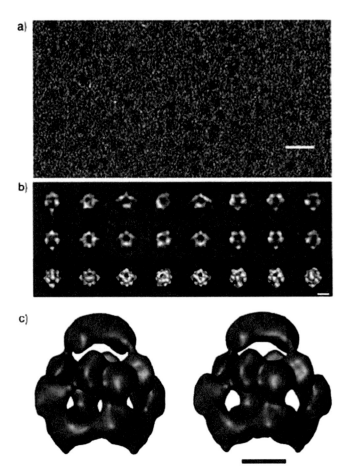

Figure 1. (a) Representative electron micrograph of calixarene micelles (the bar is 100 Å). **(b)** Row 1 shows class averages representing different spatial views of the micelles. Based on the assignment of corresponding Euler angles. 3D structure information can be retrieved at a resolution of 12 Å. Reprojections (row 2) into the 3D volume (row 3) the fit with the experimental data (bar is 50 Å). **(c)** Stereo view of the isosurface rendered 3D structure (bar is 25 Å). Reprinted with permission from ref 3. Copyright Wiley-VCH Verlag GmbH & Co. KGaA 2004.

Because the micelles are oriented randomly, the TEM-picture contains views from literally thousands of different angles. This ensemble of views can be used to calculate a 3D-structure for the micelles [10–12], as shown in Figure 1c for the micelles depicted in Figure 1.

The reconstruction shows high-density areas that were interpreted as corresponding to the carboxylate head-groups of the dendrimer and a hollow core. The low density and high mobility of the alkane chains of the dendrimer explains that the core of the micelle does not show up in the TEM picture. However, the dimensions of the micelles led to the conclusion that a relatively large concentration of water must be present in the hydrophobic core of the micelles. The first MD simulations were therefore designed to test this hypothesis and to investigate the structure of this water in a hydrophobic environment.

MOLECULAR DYNAMICS SIMULATIONS – HEPTAMERIC MICELLES

Finding a suitable starting geometry for MD simulations is always critical to their success. In this case, the 3D reconstruction of the positions of the head-groups was used to construct a putative complete micelle structure by adding the alkane chains and flooding the resulting structure with water and adding sodium ions to neutralize the ensemble. The resulting geometry was then first geometry-optimized and then equilibrated very slowly by performing successive MD simulations in which the geometrical restraints that held the micelle together were removed very slowly and carefully. A micelle resulted that was stable for 100 ns simulation time. This probably indicates that it is a stable and observable entity as unstable micelles dissociate within just a few nanoseconds in the simulations. The relaxation and equilibration led to the expulsion of the water molecules from the hydrophobic core, which became very "dry", and a concomitant shrinking of the micelle until it was approximately 5 Å smaller than suggested by the cryo-TEM images [13]. This discrepancy is larger than would normally be expected and raised the question as to exactly what the cryo-TEM images were showing.

Unusually, staining with heavy-metal derivatives was not necessary in order to be able to "see" the micelles in the TEM images. TEM is usually considered to visualize differences in density [14], so that we analyzed the density of our simulation box divided into small voxels. The results revealed areas of higher than average density associated with the sodium ions close to the anionic head-groups of the micelle. Detailed examination showed these areas of high density to correspond to contact ions pairs or ion triplets with the general structures shown in Scheme 4.

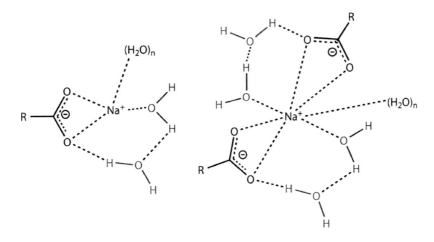

Scheme 4. Schematic structures of sodium carboxylate contact ion pairs (left) and triplets (right). Note the importance of the waters (red) that form strongly hydrogen-bonded bridges between waters (blue) coordinated directly to the sodium ion and oxygen atoms of the carboxylates.

Since the simulations were carried out with quite simple force fields that might not reproduce the behaviour of ions in aqueous solution correctly, we tested their stability by using snapshots from PM3 [15, 16] MD simulations of sodium formate in water as starting structures for geometry optimizations using density-functional theory (DFT). These calculations confirmed that structures of the types shown in Figure 2 are both stable and persistent in simulations and on geometry optimization.

This observation resolves the apparent difference between the stable micelle structures found in the simulations and the 3D-reconstructions from the cryo-TEM images. The carboxylate head-groups and their associated sodium ions together provide the high-density regions that appear in the unstained cryo-TEM images. This observation has since been confirmed for several systems in which the micelles are not observable in the TEM without staining if potassium, rather than sodium counterions are present. Figure 2 shows the time-averaged regions of high density, which should correspond to the 3D-reconstruction obtained from the cryo-TEM images from a simulation with sodium ions compared to a simulation with potassium ions only.

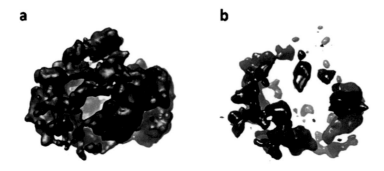

Figure 2. Snapshots taken from micelle simulations with sodium **(a)** and potassium **(b)** counterions. The figures show isodensity surfaces at a density value 20% higher than the mean for the snapshot. Reprinted with permission from ref 13. Copyright Wiley-VCH Verlag GmbH & Co. KGaA 2009.

One further aspect of the simulations [13] was also noteworthy; it proved far more difficult to obtain a stable micelle structure using the procedure described above if potassium, rather than sodium ions were used. This effect was traced to a far larger concentration of sodium ions than potassium in the immediate environment of the polycarboxylate head-groups of the dendrimers. Figure 3 shows an analysis of the time-averaged concentration of alkali-metal ions in simulations of the micelle with sodium and potassium counterions. The sodium ions associate far more tightly with the dendrimer head-groups and remain associated for far longer than their potassium counterparts [13]. This specific counterion effect was found to be responsible for the higher stability of micelles in 5:1 sodium:potassium buffer than with only potassium ions.

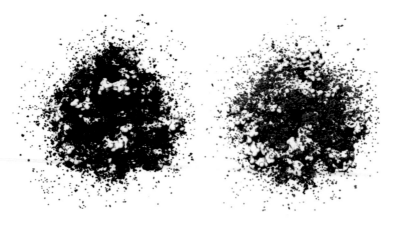

Figure 3. Areas of high sodium (blue) and potassium (green) ion density around the surface of the micelle (white).

SPECIFIC ION EFFECTS AND THE HOFMEISTER SERIES

Franz Hofmeister was born in Prague in 1850 and died in Würzburg in 1922 [17]. He studied medicine in Prague and became Professor of Pharmacology there in 1885. After Czech became the only language at the Charles University in Prague, he moved to Strasbourg in 1896, but was eventually forced to move to Würzburg when Strasbourg became French. He enjoyed a remarkable career and was the first to suggest that peptides and proteins consist of amino-acid residues connected by amide bonds. This honour is often accorded Emil Fischer, but Hofmeister spoke before Fischer at the conference in which both announced their discovery. Hofmeister is best known for what is now known as the Hofmeister series [18 – 21]. This series now describes the effects of ions on the solubility of proteins in water, although Hofmeister never formulated it in terms of individual ions, but rather for salts. Although phenomenological rationalizations for the Hofmeister series abound [22 – 26], no really convincing microscopic explanation exists. Early ideas that ions could provoke (or destroy) long-range order in water proved not to be correct [18, 27 – 29]. However, the effects described by the Hofmeister series clearly affect self-aggregation by polyelectrolytes and may therefore even determine the shape, size and stability of structurally persistent micelles.

This sensitivity of structurally persistent micelles to counterion effects makes them ideal research objects for investigating Hofmeister-like effects on polyelectrolytes as the micelles are uniformly structured and react strongly to changes in their ionic environment. These sensitive but nonetheless well defined systems provide an unprecedented level of information about specific ions effects in general and also about the factors that affect micelle structure and stability.

Experimental (cryo-TEM) tests of the differences between sodium and potassium counterions on the structures and stability of the micelles revealed strong effects. Replacing the original 5:1 Na:K buffer with a pure potassium one at the same concentration resulted in larger micelles than those observed originally. Remarkably, the original heptameric micelles were obtained by titrating the solution with a five-fold excess of pure sodium buffer [13]. At higher concentrations, the Na:K buffer gave the original heptameric micelles once more, but the cryo-TEM images revealed fewer than at lower concentrations. At the same high concentration, a pure potassium buffer gave no micelles at all. Figure 4 shows the relevant cryo-TEM images.

100

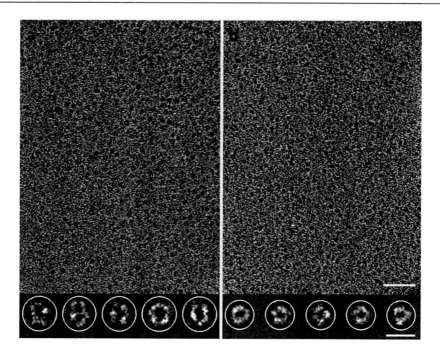

Figure 4. Cryo–TEM images of **1** in the presence of 0.027 M potassium **(a)** and sodium/potassium (5:1) **(b)** phosphate buffer. The images reveal significant differences in the micelle diameter (~10 vs. ~7 nm). Each of the above images is combined with a row of selected class sum images calculated from corresponding data sets of 2,300 images of individual assemblies. The general difference in the diameter between the two preparations is obvious and highlighted by white circles. Scale bars are 500 Å **(above)** and 100 Å **(below)**. Reprinted with permission from ref 13. Copyright Wiley-VCH Verlag GmbH & Co. KGaA 2009.

These results not only emphasize the importance of specific counterion effects, but also suggest a possible rationalization for the experimental observations. The heptameric micelles are marginally stable in potassium solution because the binding provided by the interaction of the hydrophobic chains in their core is just large enough to hold them together. The additional stabilization provided by sodium counterions stabilizes the small micelles. This interpretation is consistent with the observation [30] that ultrasonification of the original solution of **1** (with Na:K buffer) under a layer of hexane for 24 hours results in larger (dodecameric) micelles, once again with a well defined and persistent structure. The original heptameric micelles remained unchanged when treated with ultrasound for 24 hours without a hexane layer, showing that the change is caused by the hexane.

Once again, MD simulations were used to test the stability of micelles constructed on the basis of the TEM images. The fact that "unstable" micelle structures dissociate within a few nanoseconds in the simulations allowed us to study many possible starting structures with varying amounts of hexane in the core of the micelle. All simulations except those with 36 hexane molecules led to fast dissociation of the micelles into smaller aggregates, whereas that with 36 remained stable over 100 ns, both with sodium and potassium counterions [30]. The structure of the micelles in the simulations consisted of an equatorial ring of seven dendrimers with two different caps of two and three, whereas the reconstructed cryo-TEM 3D structures suggest two equivalent caps, each consisting of three dendrimer monomers, and a central ring of six dendrimers. This discrepancy is small and may either be caused by force-field deficiencies or by the fact that the MD simulations sampled a slightly less stable structure than that found in the experimental studies. However, the extremely well resolved cryo-TEM images pointed to a further indication of the importance of the alkali-metal counterions.

ATTRACTION BETWEEN POLYCARBOXYLATES

The 3D-reconstructions of the cryo-TEM images suggest orientations for the polycarboxylate head groups that indicate them to be attracting each other, as shown in Figure 5.

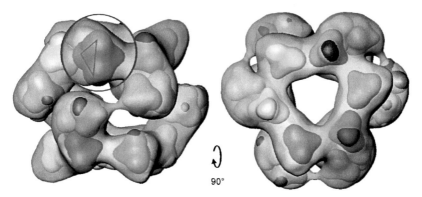

90°

Figure 5. 3D-reconstruction of the micelles of **1** after 24 hours ultrasonification with n-hexane. Reconstructed volume with visually fitted dendron fragments of **1** simplified to tetrahedra to represent their overall t-butyl-like shape (tetrahedra of the same color belong to the same molecule, the calixarene ring and alkyl chains are not shown for clarity). The circled area on the left shows a magnification of the three branches of a single dendron (highlighted by the red triangle). Reprinted in part with permission from ref 30. Copyright 2010 American Chemical Society.

Closer examination of the results of the MD simulations [30] revealed that the micelles formed by the sodium salt have a more compact and less dynamic structure than their potassium equivalent. Closer examination revealed that the orientation of the polycarboxylate head groups in the former case does indicate attraction between head groups of different dendrimer monomers and that this attraction is caused by bridging sodium ions in $RCO_2^-...Na^+...^-O_2CR$ ion triplets and quadruplets. A snapshot of the interstitial area between two head groups is shown in Figure 6.

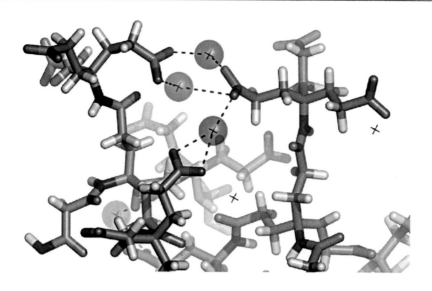

Figure 6. Snapshot of the interstitial area between two headgroups of the micelle. The dendrimer head-groups are from two different dendrimers, which are colored blue and orange. Three bridging sodium ions (blue spheres) are shown that bind three carboxylate groups together. Two non-bridging sodium ions are shown as blue crosses. Reprinted with permission from ref 30. Copyright 2010 American Chemical Society.

These results suggest a counterintuitive attraction between the polyanionic head groups. This effect was confirmed by further cryo-TEM images, in which the micelles were observed to form hexagonal "superlattices" in solution with sodium buffer, but not with potassium [31]. These structures represent a further important step in the controlled self-organization of soft particles as they provide an additional level of organization over and above that represented by the micelles themselves. Further experimental studies were based on a new single-tailed Newkome-dendritic surfactant **4** that lacks a central linking unit and varies in the length of the hydrophobic alkane chain. In this study [31], the effect of counterions on the formation of such structurally persistent micelles was investigated systematically for the first time. The critical micelle concentrations (CMCs) were found to decrease with increasing alkyl chain length, but most importantly were lower for the pure sodium salts than either for lithium or potassium, confirming the special role of sodium counterions in stabilizing polycarboxylate micelles.

MD simulations designed specifically to investigate the "head to head" aggregation of polycarboxylate dendrimers confirmed that sodium counterions can mediate an attractive interaction between deprotonated polycarboxylates by taking up bridging positions between carboxylate anions. The pH-dependence of micelle formation [31] suggests that these Na^+ bridges stabilize more strongly than conventional hydrogen bonds between protonated carboxylic acids.

One remarkable feature of all these results is how well the simple force-field based simulations are able to reproduce the quite subtle effects observed experimentally. There has been considerable discussion of the lack of accuracy of force fields for metal ions, or more accurately of combinations of force fields for metal ions and for water [32 – 34]. Surprisingly, the interactions between ions of opposite charge seem to be far less critical than the hydration of ions. Both direct MD simulations using the PM3 semi-empirical molecular orbital Hamiltonian and subsequent geometry optimizations with density-functional theory density reproduce the structures [31] observed in snapshots from the classical MD simulations. The strongest argument for the reliability of the simulations, however, is that they have been able to reproduce and in many cases even predict the unusual effects observed experimentally.

Summary and Outlook

Structurally persistent micelles remain fascinating research objects, both from the point of view of potential technological applications and because they reveal effects that are hidden in more complex or less well defined systems. Above all, the combination of synthesis, cryo-TEM and simulations has proven to be extraordinarily powerful and to lead to significant progress. It is important in this respect that impulses for new research directions and specific experiments or simulations may come from both experiment and simulation.

Acknowledgments

We are especially grateful for support from the Interdisciplinary Center for Molecular Materials (ICMM) of the Universität Erlangen-Nürnberg, from the Excellence Cluster "*Engineering of Advanced Materials*" (EAM), funded by the Deutsche Forschungsgemeinschaft and a generous funding to C. B. (BO 1000/6 – 1).

References

[1]   Shah, D.O. (Ed.) *Micelles, Microemeulsions and Monolayers*, Marcel Dekker, New York, 1998.

[2]   IUPAC. Compendium of Chemical Terminology, 2nd ed. (the "Gold Book"). Compiled by McNaught, A.D. and Wilkinson, A. Blackwell Scientific Publications, Oxford (1997). XML on-line corrected version: http://goldbook.iupac.org (2006) created by Nic, M., Jirat, J., Kosata, B. updates compiled by Jenkins, A. ISBN 0 – 9678550 – 9-8.
       doi: 10.1351/goldbook.

[3]   Kellermann, M., Bauer, W., Hirsch, A., Schade, B., Ludwig, K., Böttcher, C. (2004) The First Account of a Structurally Persistent Micelle. *Angew. Chem. Int. Ed.* **43**:2959–2962.
doi: 10.1002/anie.200353510

[4]   Burghardt, S., Hirsch, A., Schade, B., Ludwig, K., Böttcher, C. (2005) Switchable Supramolecular Organization of Structurally Defined Micelles Based on an Amphiphilic Fullerene. *Angew. Chem. Int. Ed.* **44**:2976–2979.
doi: 10.1002/anie.200462465

[5]   Schade, B., Ludwig, K., Böttcher, C., Hartnagel, U., Hirsch, A. (2007) Supramolecular Structure of 5-nm Spherical Micelles with D_3 Symmetry Assembled from Amphiphilic [3:3]-Hexakis Adducts. *Angew. Chem. Int. Ed.* **46**:4393–4396.

[6]   Hirsch, A. (2008) Amphiphilic architectures based on fullerene and calixarene platforms: From buckysomes toshape-persistent micelles. *Pure Appl. Chem.* **80**:571–587.
doi: 10.1351/pac200880030571

[7]   Becherer, M., Schade, B., Böttcher, C., Hirsch, A. (2009) Supramolecular Assembly of Self-Labeled Amphicalixarenes. *Chem. Eur. J.* **15**:1637–1648.
doi: 10.1002/chem.200802008

[8]   Schmidt, C.D., Böttcher, C., Hirsch, A. (2007) Synthesis and aggregation properties of water-soluble Newkome-dendronized perylenetetracarboxdiimines. *Eur. J. Org. Chem.* 5497–5505.

[9]   Schmidt, C.D., Böttcher, C., Hirsch, A. (2009) Chiral Water-Soluble Perylenediimides. *Eur. J. Org. Chem.* 5337–5349.
doi: 10.1002/ejoc.200900777

[10]   van Heel, M., Harauz, G., Orlova, E.V., Schmidt, R., Schatz, M. (1996) A new generation of the IMAGIC image processing system. *J. Struct. Biol.* **116**:17–24.
doi: 10.1006/jsbi.1996.0004

[11]   van Heel, M. (1987) Angular reconstitution: a posteriori assignment of projection directions for 3D reconstruction. *Ultramicroscopy* **21**:111–123.
doi: 10.1016/0304-3991(87)90078-7

[12]   Orlova, E.V., Dube, P., Harris, J.R., Beckman, E., Zemlin, F., Markl, J., van Heel, M. (1997) Structure of keyhole limpet hemocyanin type 1 (KLH1) at 15 A resolution by electron cryomicroscopy and angular reconstitution. *J. Mol. Biol.* **271**:417–437.
doi: 10.1006/jmbi.1997.1182

[13] Jäger, C.M., Hirsch, A., Schade, B., Böttcher, C., Clark, T. (2009) Counterions Control the Self-Assembly of Structurally Persistent Micelles; Theoretical Prediction and Experimental Observation of Stabilization by Sodium Ions. *Chem. Eur. J.* **15**:8586 – 8592.
doi: 10.1002/chem.200900885

[14] Williams, D.B., Carter, C.B. *Transmission Electron Microscopy: A Textbook for Materials Science*, Plenum Press, New York, 1996.

[15] Stewart, J.J.P. (1989) Optimization of parameters for semiempirical methods I. Method. *J. Comp. Chem.* **10**:209 – 220.
doi: 10.1002/jcc.540100208

[16] Stewart, J.J.P. (1989) Optimization of parameters for semiempirical methods II. Applications. *J. Comp. Chem.* **10**:221 – 246.
doi: 10.1002/jcc.540100209

[17] Abernethy, J.L. (1967) Franz Hofmeister – the Impact of his Life and Research on Chemistry. *J. Chem. Ed.* **44**:177 – 180.
doi: 10.1021/ed044p177

[18] See, for instance Zhang, Y., Cremer, P.S. (2006) Interactions between macromolecules and ions: The Hofmeister series. *Curr. Op. Chem. Biol.* **10**:658-663.

[19] Hofmeister, F. (1888) Zur Lehre von der Wirkung der Salze. Zweite Mittheilung *Arch. Exp. Pathol. Pharmakol.* **24**:247 – 260.
*doi:*10.1007/BF01918191

[20] Kunz, W., Henle, J., Ninham, B.W. (2004) Zur Lehre von der Wirkung der Salze (About the science of the effect of salts): Franz Hofmeister's historical papers. *Curr. Op. Colloid Interface Sci.* **9**:19 – 37.
doi: 10.1016/j.cocis.2004.05.005

[21] Kunz, W., LoNostro, P., Ninham, B.W. (2004) The Present State of Affairs with Hofmeister Effects. *Curr. Op. Colloid Interface Sci.* **9**:1 – 18.
doi: 10.1016/j.cocis.2004.05.004

[22] Collins, K.D, Washabaugh, M.W. (1985) The Hofmeister effect and the behaviour of water at interfaces. *Q. Rev. Biophys.* **18**:323 – 422.
doi: 10.1017/S0033583500005369

[23] Washabaugh, M.W., Collins, K.D. (1986) The systematic characterization by aqueous column chromatography of solutes which affect protein stability. *J. Biol. Chem.* **261**:12477 – 12485.

[24]   Collins, K.D. (1995) Sticky ions in biological systems. *Proc. Natl. Acad. Sci. U.S.A.* **92**:5553 – 5557.
doi: 10.1073/pnas.92.12.5553

[25]   Neilson, G.W., Enderby, J.E. (1996) Aqueous Solutions and Neutron Scattering. *J. Phys. Chem.* **100**:1317 – 1322.
doi: 10.1021/jp951490y

[26]   Enderby, J.E. (1995) Ion solvation via neutron scattering. *Chem. Soc. Rev.* **24**:159 – 168.
doi: 10.1039/cs9952400159

[27]   Collins, K.D., Neilson, G.W., Enderby, J.E. (2007) Ions in water: characterizing the forces that control chemical processes and biological structure. *Biophys. Chem.* **128**:95 – 104.
doi: 10.1016/j.bpc.2007.03.009

[28]   Marcus, Y. (2009) Effect of ions on the structure of water: Structure making and breaking. *Chem. Rev.* **109**:1346 – 1370.
doi: 10.1021/cr8003828

[29]   Collins, K.D. (2006) Ion hydration: Implications for cellular function, polyelectrolytes, and protein crystallization. *Biophys. Chem.* **119**:271 – 281.
doi: 10.1016/j.bpc.2005.08.010

[30]   Jäger, C.M., Hirsch, A., Schade, B., Ludwig, K., Böttcher, C., Clark, T. (2010) Self-Assembly of Structurally Persistent Micelles is Controlled by Specific Ion Effects and Hydrophobic Guest. *Langmuir* **26**:10460 – 10466.

[31]   Rosenlehner, K., Schade, B., Böttcher, C., Jäger, C.M., Clark, T., Hirsch A. (2010) Sodium-Effect on the Self-Organization of Amphiphilic Carboxylates: Formation of Structured Micelles and Superlattices. *Chem. Eur. J.* **16**:9544 – 9554.
doi: 10.1002/chem.201001150

[32]   Joung, I.S., Cheatham, III, T.E. (2008) Determination of alkali and halide monovalent ion parameters for use in explicitly solvated biomolecular simulations. *J. Phys. Chem.* **112**:9020 – 9041.
doi: 10.1021/jp8001614

[33]   Horinek, D., Mamatkulov, S.I., Netz, R.R., (2009) Rational design of ion force fields based on thermodynamic solvation properties. *J. Chem. Phys.* **130**:124507.
doi: 10.1063/1.3081142

[34]   Hess, B., van der Vegt, N. (2009) Cation specific binding with protein surface charges. *Proc. Natl. Acad. Sci. U.S.A.* **109**:13296 – 13300.
doi: 10.1073/pnas.0902904106l

ENERGY TRANSDUCTIONS IN ATP SYNTHASE

CHRISTOPH VON BALLMOOS, ALEXANDER WIEDENMANN AND PETER DIMROTH*

Institute of Microbiology, Swiss Federal Institute of Technology (ETH) Zürich, Wolfgang-Pauli-Straße 10, 8093 Zürich, Switzerland

E-MAIL: *dimroth@micro.biol.ethz.ch

Received: 30th August 2010 / Published: 13th June 2011

ABSTRACT

Life depends on chemical energy in the form of adenosinetriphosphate (ATP), most of which is produced by the F_1F_o ATP synthase, a rotary nanomachine. Here, we report new insights into the torque-generating mechanism by the membrane-embedded F_o motor. High coupling ion concentrations on the ion entrance side are required to prepare the motor for rotation. In the absence of this condition, the motor is in a resting state, where the stator arginine is assumed to form a complex with the conserved carboxyl group of an empty rotor site. At high coupling ion concentrations, however, the motor switches into a mobile state, in which an incoming coupling ion has displaced the arginine and has formed a new complex with the rotor site. At appropriate driving forces, the ion dissociates from the next incoming rotor site and escapes into the cytoplasmic reservoir of the membrane. The resulting negatively charged site is attracted by the positively charged arginine, which elicits rotation and generates torque.

INTRODUCTION

Life is dependent on a continuous input of energy. Light or nutrients serve as external energy source which is converted into the cell intrinsic energy currency ATP. In order to do so, photosynthetic reaction centres in specific membranes of green plants or phototrophic bacteria harvest the light to transfer electrons and protons across the membrane and thereby generate an electro-chemical proton gradient. Animals and heterotrophic bacteria use energy from the combustion of nutrients for proton translocation by respiratory enzymes to generate an electrical potential across the membrane, while certain anaerobic bacteria dispose of membrane-bound Na^+-translocating decarboxylases to carry out this task [1]. The free energy thus stored in the charged membranes is used by the universal F_1F_o ATP synthase to combine ADP and phosphate to ATP, thereby conserving the energy in its terminal pyrophosphate bond (Figure 1). Upon hydrolysis of this bond, the stored energy is released and can be used to drive numerous energy-consuming cellular reactions, *e.g.* biosynthesis of many cellular compounds, ion transport, motility, signal transduction, etc.. Hence, continuous cycling between ATP synthesis and hydrolysis is a prerequisite to fulfil the cell's energy needs. The dimension of this cell energy cycle is astounding, amounting to a daily turnover of 50 kg ATP in a human on average.

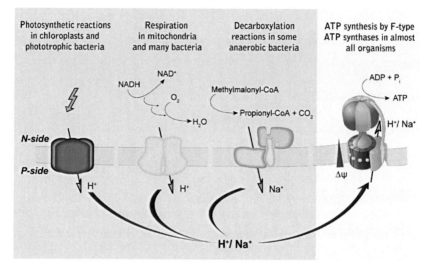

Figure 1. Ion cycling across biological membranes leading to ATP synthesis (modified from [5]). Chloroplasts and phototrophic bacteria convert light energy into an electrochemical proton gradient across the membrane. In heterotrophic organisms like animals or bacteria the metabolism of nutrients generates reducing equivalents (NADH, succinate) that are reoxidized by the respiratory chain enzymes using oxygen as terminal electron acceptor. The free energy is converted into an electrochemical gradient of protons across the membrane. The anaerobic bacterium *P. modestum* couples the decarboxylation of methylmalonyl-CoA to electrogenic Na^+-transport. The electrochemical H^+ or Na^+ gradients established by these membrane-bound complexes serve as energy source for ATP synthesis from ADP and inorganic phosphate by an F_1F_o ATP synthase.

The ATP synthase is a miniature machine composed of two rotary motors, F_1 and F_o, which are mechanically connected to assure energy exchange amongst them [2 – 5]. The F_1 motor is a water soluble protein complex which displays the following subunit composition α_3, β_3, γ. δ, ε. The F_o motor is membrane-embedded and exists in its simplest form as subunits a, b_2, and c_{10-15} (Figure 2). In the ATP synthesis mode, the F_o motor converts the electrochemical gradient of H^+ or Na^+ ions into torque, enabling the F_1 motor to act as an ATP generator. In the ATP hydrolysis mode, F_1 converts the chemical energy of ATP hydrolysis into torque, causing the membrane-embedded F_o motor to act as an ion pump. Generally, the F_o motor generates the larger torque and forces the F_1 motor to function in ATP synthesis direction. In most eukaryotes and some bacteria, the hydrolysis mode is tightly regulated or blocked to avoid unnecessary energy waste. However, in fermenting bacteria, when the respiratory enzymes are not active, the F_1 motor hydrolyses ATP to employ the F_o motor to generate the membrane potential which is indispensable for every living cell.

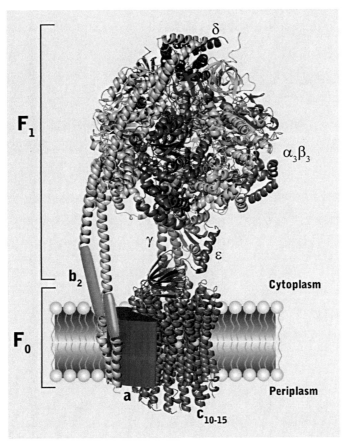

Figure 2. Organization of a bacterial F_1F_o ATP synthase (modified from [5]). The structures of individual sub-complexes were taken from RCSB Protein Data Base and assembled by eye according to biochemical data. The structures used were from the c ring of *I. tartaricus* (1CYE), the F_1 organization of *E. coli* (1JNV), the δ subunit of *E. coli* (2A7U), the peripheral stalk from bovine mitochondria (2CLY) and the membranous part of subunit b of *E. coli* (1B9U). No high resolution structural data are available for subunit a and the hinge region of subunit b. The picture was created using Pymol (DeLano Scientific LLC).

The crystal structure of F_1 from bovine heart mitochondria revealed a hexameric assembly of alternating α and β subunits around an eccentric α-helical coiled-coil γ subunit [6]. The catalytic centre consists of three nucleotide binding sites on the β subunits at the interface with the α subunits. At any one time each of these sites is in a different conformation and performs a different task in the synthesis or hydrolysis of ATP. The actual conformation of a site depends on its specific interaction with the γ subunit which changes upon rotation of γ in a coordinated sequence. Each site thus cycles through the same set of open and closed conformations with the corresponding functions in the uptake, processing and release of nucleotides [7].

Kinetic details of the binding change mechanism have been revealed from single molecule studies with isolated F_1 molecules. Thereby, the γ subunit was observed to rotate in discrete steps of $120°$ for each ATP molecule hydrolyzed [8]. Under limiting ATP concentrations, the $120°$ rotation can be divided into $80°$ and $40°$ sub-step rotations, associated with ATP binding and ATP hydrolysis followed by product release, respectively [9, 10]. Single molecule experiments were also performed with the intact F_1F_o ATP synthase reconstituted in lipid vesicles [11, 12]. Here, compiled by the subunits γ, ε and the $c_{10\text{-}15}$ ring the rotor turns counter clockwise in the ATP hydrolysis direction and clockwise in the ATP synthesis direction, when viewed from the F_o domain. When proteoliposomes prepared with F_1F_o from *Escherichia coli* were properly energized for ATP synthesis, the rotor turned with a stepping size of $36°$, reflecting a stepping motion for each individual c subunit in the oligomeric c_{10} ring [13].

THE F_o MOTOR

The rotary F_o motor of the ATP synthase obtains its driving force from an electrochemical ion gradient across the membrane. The only other ion-driven rotary motors in nature are the A_o portion of archaeal A_1A_o ATPases, the V_o portion of vacuolar V_1V_o ATPases and the flagellar motor. While A_o and V_o are phylogenetically related to F_o, the flagellar motor is much larger and has a different construction. The central element of the F_o motor is an oligomeric ring of $10-15$ copies of c subunits. Each c subunit consists of two membrane-spanning α-helices that are connected by a hydrophilic loop on the cytoplasmic side of the membrane. In the c ring, the loop regions form an extensive area of contact, where the ring combines with the foot of the central stalk subunits γ and ε to form the rotor assembly [14, 15]. At the periphery of the c ring, the F_o motor consists of subunits a and b_2. Subunit a is a very hydrophobic protein containing 5 or 6 membrane-spanning α-helices, which actively participates in the ion translocation mechanism [16, 17]. The b subunits are anchored within the membrane by an N-terminal α-helix and extend as the peripheral stalk all the way to the head of the F_1 domain [18-20]. Near their N-terminus, the b subunits contact the C-terminal part of the c subunits and the periplasmic loop of subunit a between helices 4 and 5. With their C-terminal ends, the b subunits form a strong complex with the δ and the α subunit and thus serve to connect the stator parts of F_o (a, b_2) with those of F_1 (α_3, β_3, δ). This connection is an important device to counter the tendency of the stator to follow the rotation of the rotor. The X-ray structure of the mitochondrial peripheral stalk indicates that it is rather rigid and therefore well suited for its anticipated function [21].

STRUCTURE OF C RINGS AND THEIR BINDING SITES

Na⁺ binding sites

Na⁺ binding sites

For a functional understanding of the F_o motor, structural knowledge is essential. Hence, the high resolution structure of the c ring from the Na⁺-translocating ATP synthase of *Ilyobacter tartaricus* was an important step towards this goal. In the X-ray structure at 2.8 Å resolution [22], the oligomeric ring with 11 subunits appears as an hourglass-shaped hollow cylinder (Figure 2). The centre of the cylinder encloses a hydrophobic cavity which is filled with phospholipids in the natural environment of the membrane [23]. Each c monomer folds into a helical hairpin with the loop at the cytoplasmic side and the termini at the periplasmic side of the membrane. The N-terminal helices form a tightly packed inner ring with no space for large side chains which is accounted for by a conserved motif of four glycine residues [24]. In addition, the tight packing of the glycines causes the inner ring of helices to narrow from the cytoplasm toward the middle of the membrane. The 11 C-terminal helices of the outer ring pack into the grooves of the inner ring in the cytoplasmic half of the protein. All helices show a bend of about 20° in the middle of the membrane (at P28 and E65 in the N-terminal and C-terminal helices, respectively), marking the narrowest part of the hourglass shape. Eleven Na⁺ ions are located here at their binding sites, facing toward the outer surface of the c ring, which is consistent with cross-linking data [25].

With respect to the function, the most important insights derive from the structure of the ion binding site. Each of the 11 Na⁺ ions is bound at the interface of one N-terminal and two C-terminal helices. The coordination sphere is formed by the side-chain oxygens of Q32 and E65 of one subunit and of the side chain oxygen of S66 plus the carbonyl oxygen of V63 of the neighbouring subunit. A fifth Na⁺ coordination site is provided by the oxygen of a structural water molecule, which is held in place by donating hydrogen bonds to the carbonyl oxygen of A64 and the side chain oxygen of T67 [26] (Figure 3a). The structure of the ion binding site is remarkably similar to that of the larger k-ring from the V-ATPase of *Enterococcus hirae* [27]. In this structure, all five Na⁺ coordination sites are directly contributed from the protein and structural water is not involved. The coordination sphere is completed by a network of hydrogen bonds. The side chain oxygen of E65 which is involved in Na⁺ coordination also receives a hydrogen bond from the NH₂ group of Q32, and the other oxygen of E65 receives hydrogen bonds from the OH groups of S66 and Y70. These hydrogen bonds serve to keep E65 deprotonated at physiological pH in order to allow Na⁺ binding. The mutation Y70F, *e.g.*, prohibits hydrogen bonding of Y70 with E65 and leads to a 30-fold decreased affinity for Na⁺ [31]. The wild type arrangement, however, creates a stable, locked conformation, from which horizontal transfer of the Na⁺ ion to subunit a is prevented. In the subunit a/c interface, the conformation of the binding site must therefore convert to an open one, to allow ion transfer between the binding site and subunit a (see below).

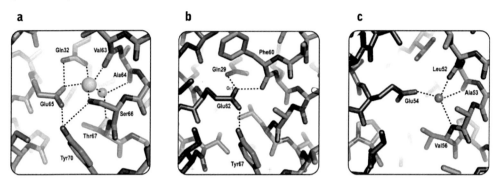

Figure 3. View on the three types of ion binding sites in F-type ATP synthases. (a) $Na^+ + H_2O$ ion binding site in the c_{11}-ring of *I. tartaricus* (2wgm). **(b)** H^+-binding site in the c_{15}-ring of *S. platensis* (2wie). **(c)** $H^+ + H_2O$-binding in the c_{13}-ring of *B. pseudofirmus* OF4 (2 × 2v). Every binding site is made up from residues of two neighbouring subunits (brown and cyan). Amino acids involved in ion binding site coordination are indicated. The hydrogen bonding network is indicated by dashed lines. The Na^+ ion (yellow) and the water molecules (red) are shown as spheres. In **(b)** The oxygen atom on Glu62 harbouring the H^+ is highlighted with $O_{\varepsilon 1}$.

Two different ion binding modes can be envisaged for the proton-translocating ATP synthases. Either, the binding site involves several ligands to form a coordination sphere for ion binding in analogy to that of the Na^+-translocating ATP synthases. In this case, the proton would be bound to water and the resulting hydronium ion would be coordinated. Alternatively, the entire binding site could consist of just the conserved carboxylate allowing protons to hop on and off in a group protonation type mechanism, as proposed in most F_o motor models. The structure of the c_{15} ring from the H^+-translocating ATP synthase from the cyanobacterium *Spirulina platensis* [28] shows that its overall shape resembles that of the c ring from *I. tartaricus*. Intriguingly, the conformation of the conserved glutamate that is essential for ion binding is stabilized by a hydrogen bonding network which is similar to that of the Na^+ binding c-rings. As the authors point out, the conserved glutamate must be protonated in the crystallization buffer at pH 4.3 and indeed, one oxygen atom ($O_{\varepsilon 1}$) of this glutamate appears to act as a hydrogen bond donor for the backbone carbonyl oxygen of F60 from the neighbouring subunit (Figure 3b). Hence, the pH 4.3 structure is in accordance to the group protonated carboxylate that is expected at this acidic pH. However, it does not provide any clues on whether the binding site glutamate remains in its group protonated form at physiological pH values or participates in the coordination of a hydronium ion. To discriminate between these options, the structure must be determined at neutral pH.

Very recently, the c-ring structure of the alkaliphilic bacterium bacillus OF4 has been determined from crystals grown at pH 4.3 and a water molecule is clearly visible in the ion binding pocket (Figure 3c) [29]. Furthermore, the F1-c10 structure from yeast suggests the coordination of water because H-bonding as described above for the *S. platensis* c-ring is not possible [30]. Interestingly, the yeast structure has been determined from crystals grown at pH 7.5. Although it cannot be clarified from the existing binding site structures, if the coupling proton is indeed coordinated as a hydronium ion, the presence of structured water renders old models of simple protonation/deprotonation of the ion binding carboxylates highly unlikely.

Besides this structural work, the mode of proton binding has been investigated by biochemical techniques [31]. The well known modification of the binding site carboxylate with the inhibitor *N,N'*-dicyclohexylcarbodiimide (DCCD) consumes a proton and can therefore be used to monitor the proton availability at the binding site. The pH profile for DCCD labelling of the ATP synthase from *Halobacterium salinarium* or *I. tartaricus* in its H^+-translocation mode (in the absence of Na^+) showed a typical titration curve with maximum labelling rates at pH 6 and below, half maximal labelling rates around pH 7, and no detectable labelling at pH 8 and above. Hence, these enzymes show the expected behaviour for a group protonation of the binding site carboxylate.

For the majority of the H^+-ATP synthases, including enzymes from *E. coli*, spinach chloroplasts and bovine mitochondria, however, bell-shaped labelling pH profiles were observed with maximum labelling rates between pH 7.5 and 9 and decreasing rates at lower or higher pH values. These data are clearly not compatible with a simple group protonation of the binding site. They would be consistent, however, with hydronium ion binding in the physiological (neutral) pH range and a shift to group protonation at acidic pH, when the proton concentration approaches the dissociation constant of the binding site carboxylate. The decreasing labelling rates at acidic pH values may be related to the robust conformation of the binding site seen in the group-protonated structure of the *S. platensis* c ring at pH 4.3 [28]. This may have an inferior reactivity for substitution by DCCD compared to the conformation with a bound hydronium ion in the physiological near neutral pH range. Higher labelling rates at pH 7 than at pH 5.5 or 10 were also observed with the *S. platensis* c ring, but whether this can be attributed to structural differences of the binding pocket at the different pH values has not yet been rationalized [32].

These results not only provide evidence for distinct proton binding modes in the ATP synthase family but also show that the degree of binding site protonation depends on the external pH. As DCCD reacts with binding sites outside the subunit a/c interface, this implies that there is an access pathway for the protons to the binding sites in the absence of subunit a [33, 34]. Whether this access route is sufficiently fast to meet physiological conditions is presently unknown. Multiple lines of evidence indicate that the ions reach the sites from the cytoplasmic reservoir of the membrane. Interestingly, the pH profiles for

DCCD labelling of the various ATP synthases were remarkably similar to those of the ATP hydrolysis activities of the F_1F_o holoenzymes [31]. We anticipate that the rate limiting step in the ATP hydrolysis mode of the enzyme is the loading of the binding site in the subunit a/ c interface from the cytoplasmic side of the membrane, because electrostatic constraints prevent the rotation of an unoccupied site with its negative charge into the hydrophobic membrane environment [35]. Hence, the rate of ATP hydrolysis is dependent on the cytoplasmic pH value and on the dissociation constant of the binding site at the cytoplasmic access route in the subunit a/c interface. Similar dissociation constants for the site at this location and outside the interface with subunit a may be the rationale for the similarity of the DCCD labelling and ATP hydrolysis pH profiles.

THE ROLE OF SUBUNIT A

Structural information on subunit a is sadly very limited, but important details of its function have nevertheless been uncovered in the past decades. There is general agreement that the a subunit contributes the physiological access routes from the periplasm and the cytoplasm to the binding sites and that the conserved arginine is located in between these routes, blocking unhindered proton flux. In elaborative studies with cysteine mutants of subunit a from *E. coli*, the ion accessible parts have been probed [36, 37]. These data suggest that an ion accessible channel, located in the centre of helices 2–5 of subunit a, extends from the periplasm to the centre of the membrane. The cytoplasmic access route to the binding site, however, remained ambiguous, because only few residues near the cytoplasmic surface of the fourth transmembrane helix of the a subunit were accessible. Recently, the cytoplasmic half of the outer helices of subunit c were also found to contain ion accessible sites, which therefore suggests that the cytoplasmic access pathway is located in the interface between subunit a and c [38]. Such an access route would be consistent with cysteine/cysteine cross linking data with the *E. coli* and *I. tartaricus* ATP synthase, which show that helix IV of the a subunit is in close proximity to the outer helices of the c ring [39, 40]. Recent cryo-electron microscopy data on the structure of the V-ATPase of *Thermus thermophilus*, however, indicate only a small area of contact between the stator and the c ring equivalent near the centre of the membrane [41].

The most important residue of subunit a is the membrane-embedded, conserved arginine (position 210 in *E. coli*), also known as the stator charge. So far, every mutation of this residue in the *E. coli* enzyme led to a complete loss of activity [42], whereas in the Na^+-translocating *P. modestum* enzyme, mutations retaining the positive charge were tolerated under specific conditions [43]. In the ATP hydrolysis mode, the stator charge is believed to ensure dissociation of the coupling ion from a c ring site entering the interface with the a subunit. Accordingly, ATP hydrolysis became uncoupled from Na^+ translocation by mutagenesis of the stator arginine to alanine in the *P. modestum* ATP synthase. A second function of the arginine is to provide a seal between the two access pathways to prevent ion leakage across the membrane [44]. In the ATP synthesis mode, the role of the arginine is probably more sophisticated, since it not only affects the ion binding affinity of the site at the periplasmic channel, but also plays a profound role in the generation of torque (see below).

Insights into the mode of interaction of the stator arginine with the ion binding site have been obtained by cysteine-cysteine cross-linking experiments between subunits a and c of the *I. tartaricus* ATP synthase [40]. From the available c ring structure with its locked binding site conformation, one could predict a conformational change in the interface with subunit a to facilitate ion transfer between the c ring site and the a subunit. Strong evidence for a small but significant conformational change upon contact with the stator charge derives from the cross-linking patterns between cT67C or cG68C and aN230C. Whereas in the locked conformation, the two c ring residues from neighbouring subunits occupy spatially similar positions toward a potent binding partner on subunit a, only cT67C, but not cG68C, could form efficient cross-links with aN230C in the presence of the stator charge. Upon mutagenesis of the stator arginine by uncharged amino acids, the cross-linking efficiency of aN230C with cT67C remained unchanged, but that with cG68C increased significantly. From this and other data, the stator charge is proposed to induce a conformational change in the empty binding site, by which cG68C is disconnected from its cross-linking partner in subunit a. Relating these observations to the c ring structure suggests a rearrangement of the binding site residues with the side chain of cE65 being slightly pushed or rotated toward the c ring centre, thereby affecting the accessibility of cG68C (which is on the same helix) but not of cT67C (which is on the neighbouring helix). The spatial demand for this rearrangement is presumably allocated by a cavity between the inner and the outer ring of helices in the region of the two conserved residues G25 and G68 (Figure 4a). Accordingly, replacement of G25 by a bulky isoleucine increased the cross-linking efficiency of the cG68C/aN230C pair to a similar extent as the replacement of the stator charge. No ATP synthesis was observed with the cG25I mutant, emphasizing the functional importance of the small residue. The interpretations were supported by energy minimization calculations of the c ring structure with a short strip of helix 4 of subunit a, containing residues aI225 to aM231, which suggested possible modes for the interaction of the arginine with an unoccupied binding site (Figure 4b).

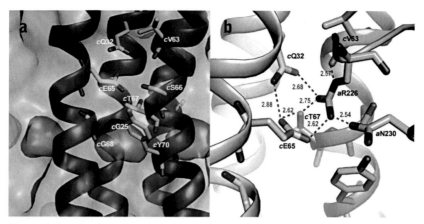

Figure 4. Interaction of the a subunit with the c-ring (modified from [40]). (a) Perspective view on the surface of the isolated c-ring of *I. tartaricus*. Shown are the atom boundaries displayed as surface to visualize the cavity below the ion binding site. Further, the residues of the ion binding site and the glycine residues cG25 and cG68 around the cavity are depicted (see text for details). **(b)** Coordination of the stator arginine after energy minimization calculations of the c-ring and a heptapeptide of helix 4 of subunit a. Depicted are the calculated positions and possible hydrogen bonds of the binding site residues on the c-ring and the stator arginine in the absence of harmonic backbone restraints of the outer helices. Putative hydrogen bond lengths are marked in Å. The figure was prepared using Pymol (DeLano Scientific).

DRIVING FORCES FOR ATP SYNTHESIS

An electrochemical gradient of either H^+ or Na^+ provides the driving force for ATP synthesis by F_1F_o ATP synthases. The downhill movement of these ions through the F_o domain generates torque which is transmitted to F_1 to energize ATP synthesis. The driving force consists of two thermodynamically equivalent parameters, the ion concentration gradient (ΔpH or ΔpNa^+) and the membrane potential ($\Delta\psi$). Most organisms use the proton motive force ($\Delta\mu H^+$) with the parameters ΔpH and $\Delta\psi$, but some anaerobic bacteria use the sodium motive force consisting of ΔpNa^+ and $\Delta\psi$ for the synthesis of ATP.

Proton motive force driven ATP synthesis was originally formulated in Mitchell's chemiosmotic model, which also implies that ΔpH and $\Delta\psi$ are equivalent kinetic driving forces [45]. Early experiments with the chloroplast enzyme seemed to support this notion, because ATP synthesis could be driven by ΔpH alone [46]. The method to establish the ΔpH by an acid/base transition with succinate as the acidic buffer, however, was later shown to also create a significant $\Delta\psi$, owing to the membrane permeability of the succinate monoanion [47]. An equivalent of ΔpH and $\Delta\psi$ as driving forces for ATP synthesis has also been described for the mitochondrial enzyme [48, 49]. Accordingly, it was predicted that ATP synthesis occurs when the thermodynamic requirements are fulfilled, i.e. if

$$\Delta\mu H^+ = \Delta\psi + \Delta pH > \Delta Gp = \Delta G^{o'}x \log_{10} [ATP]/([ADP] \times [P_i]).$$

Several investigations involving a variety of ATP synthases have subsequently shown that the equivalence of ΔpH and $\Delta\psi$ as driving force for ATP synthesis is not perfect [50 – 52], suggesting that different kinetic barriers have to be overcome by either ΔpH or $\Delta\psi$. In all these studies, except for the chloroplast enzyme from spinach, efficient ATP synthesis required a total driving force of > 180 mV, consisting of a minimal $\Delta\psi$ of ~80 mV and a corresponding ΔpH.

REQUIREMENTS FOR ATP SYNTHESIS

ATP synthesis measurements have been performed with a number of ATP synthases, but the most complete data are available for the H^+-translocating ATP synthase of *E. coli* or spinach chloroplasts and the Na^+-translocating ATP synthases of *P. modestum* or *I. tartaricus*, respectively. With *E. coli* membrane vesicles energized by respiration, a turnover number for ATP synthesis of about 270 s^{-1} was found [53]. For the reconstituted *E. coli* enzyme, the maximal rates of ATP synthesis were 80 s^{-1}, when the vesicles were energized with a total driving force of 380 mV, consisting of a ΔpH of 240 mV and a $\Delta\psi$ of 140 mV [52]. In this study, the rate decreased drastically to ~1 s^{-1} if the ΔpH of 240 mV was maintained and the $\Delta\psi$ was abolished. These results confirm the notion that ATP synthesis requires both parameters of the proton motive force and that these are kinetically not equivalent.

In the same study, ATP synthesis was also abolished by decreasing the ΔpH to ~180 mV at $\Delta\psi$ values between 110 and 150 mV. This is remarkable, because the energetic parameters ($\Delta\mu H^+ = 290 – 330$ mV) clearly exceeded those found in a growing *E. coli* cell. The discrepancy could be resolved in more recent studies with the reconstituted *E. coli* enzyme, where it was observed that the rate of ATP synthesis not only depends on the driving forces ΔpH and $\Delta\psi$, but also on the proton concentration on the source (P) side of the membrane [54]. In these experiments, efficient ATP synthesis was found at significantly lower driving forces (ΔpH = 120 mV; $\Delta\psi$ = 120 mV) than in the previous studies, if the P-side pH values were kept below 6. The rate decreased sharply at higher pH values and reached undetectable levels at pH 7 and above (Figure 5a). Effective ATP synthesis therefore not only depends on an appropriate size of the two driving forces ΔpH and $\Delta\psi$, but also on a sufficiently high proton concentration at the P-side. From a physiological point of view, this result causes a dilemma, because *E. coli* cells grow perfectly well and synthesize ATP at pH values of 8 and above. Lateral proton transfer along the membrane surface between the respiratory chain enzymes and the ATP synthase might be an elegant solution to generate sufficiently high proton concentrations at the entrance port of the ATP synthase to make ATP synthesis feasible [55]. In contrast, the requirements of the Na^+-driven ATP synthase of *P. modestum* are nicely met by its natural environment of brackish water (50 – 150 mM NaCl).

The high proton concentration requirement is not observed during ATP hydrolysis, where the fully coupled enzyme operates at nearly maximal speed at pH > 8.5. Hence, approximately 100-times higher proton concentrations are necessary to saturate the binding sites at the periplasmic entrance channel during ATP synthesis than at the cytoplasmic entrance channel during ATP hydrolysis. Studies with the ATP synthase of spinach chloroplast are in agreement with the notion above except that the enzyme is capable of synthesizing ATP in the presence of a large ΔpH only [52].

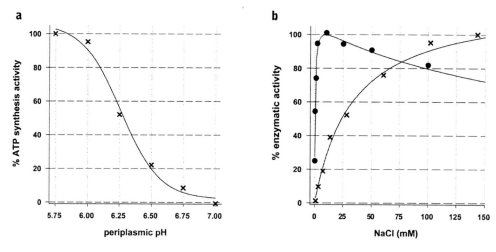

Figure 5. Importance of the coupling ion concentration on the *P*-side during ATP synthesis. **(a)** ATP synthesis with purified ATP synthase of *E. coli* reconstituted in proteoliposomes. At constant driving forces of $\Delta\psi = 120$ mV and $\Delta pH = 2$ (122 mV), half maximal ATP synthesis is observed at a P-side pH 6.2. No ATP synthesis is observed at pH values above 7. **(b)** ATP synthesis with purified ATP synthase of *P. modestum* reconstituted in proteoliposomes. At constant driving forces of $\Delta\psi = 140$ mV and $\Delta pNa = 120$ mV, half maximal ATP synthesis is observed at a P-side Na^+ concentration of ~35 mM (crosses). In contrary, half maximal ATP hydrolysis is already observed at a Na^+ concentration of ~0.5 mM (closed circles).

Corresponding observations have also been made with the Na^+-translocating ATP synthase of *P. modestum*. ATP synthesis required both parameters of the sodium motive force with a minimum ΔpNa^+ of > 30 mV and a minimum $\Delta\psi$ of > 180 mV and in addition P side Na^+ concentrations of ~35 mM for half maximal activities [54]. This Na^+ concentration requirement for ATP synthesis was approximately 70 – 100 times higher than that for half maximal ATP hydrolysis [56, 57] (Figure 5b). Hence, in spite of the different coupling ions, the ATP synthases of *E. coli* and *P. modestum* exhibit a remarkable similar asymmetry in their coupling ion concentration requirements for ATP synthesis or hydrolysis. A reasonable explanation for this asymmetry most likely reflects different affinities of the binding sites for the coupling ion at the periplasmic and cytoplasmic access routes of subunit a. The affinity of the binding site will be affected at each specific location by the chemical environment of the a subunit, particularly the interaction with the stator charge.

IMPLICATIONS FOR THE F_o MOTOR FUNCTION

In a current model of the F_o motor, the arginine is assumed to form a complex with the negatively charged empty binding site between the periplasmic and cytoplasmic access channels of subunit a [54]. Ion binding to an arginine-bound site is expected to require higher ion concentrations than ion binding to a free site. In the ATP hydrolysis mode of the enzyme, the torque applied from the F_1 motor is likely to disrupt the arginine-binding site complex with the result that the affinity of the site at the cytoplasmic channel is no longer affected by the stator charge. Very similar affinities for the binding sites outside the interface with subunit a and at the cytoplasmic entrance route during ATP hydrolysis support this view (see above). In the ATP synthesis mode of the enzyme, however, the complex of the binding site with the arginine must first be broken before the site can form the new complex with the incoming coupling ion and therefore, higher coupling ion concentrations are required to compete successfully with the stator charge to occupy the sites at the periplasmic entrance channel.

It is important that sufficiently high P-side coupling ion concentrations are an essential requirement for the synthesis of ATP, which cannot be compensated by very high $\Delta pH/\Delta pNa^+$ and/or $\Delta\psi$ values. This suggests that the events creating rotation must follow a given sequence, which starts with the replacement of the arginine by an incoming coupling ion at the periplasmic channel. Ion dissociation from the following binding site at the cytoplasmic channel is in equilibrium with the ion concentration at the cytoplasmic side and is probably favoured by the positive charge of the arginine and the membrane potential (see below). After dissociation, the new negatively charged site is immediately attracted by the positively charged arginine, causing rotation and the generation of torque and resulting in a new arginine/binding site complex.

Importantly, the membrane potential did not affect the ion binding affinity of the site of the periplasmic channel of the Na^+-ATP synthase [54]. This implies that the potential has no influence on the initial replacement of the arginine from the binding site by Na^+ entering from the periplasmic side. The potential therefore probably exerts its effect on the events at the cytoplasmic channel. This is consistent with the more hydrophobic nature of this channel compared to the periplasmic one, as observed in accessibility studies [36].

The mechanism of the F_o motor has also been investigated by ion translocation experiments through the isolated F_o part. While most of these studies did not take the orientation of F_o within the membrane into account, it is now clear that detailed insights require transport measurements in each specific direction with unidirectionally reconstituted F_o complexes [58]. Proton transport measurements were performed with the reconstituted F_o complexes of E. coli in both directions. With a constant driving force of $\Delta\psi = 73$ mV or $\Delta pH = 73$ mV the rate of proton transport was about $600\,s^{-1}$ in synthesis direction and about $1400\,s^{-1}$ in hydrolysis direction. Hence, in one specific direction, similar rates were produced by each driving force, which are therefore equivalent for the F_o part. The transport rates corresponded linearly to the applied driving forces and no voltage threshold was apparent. The most significant difference between H^+ transport in synthesis and hydrolysis direction was observed, when the influence of the pH was determined. Proton transport in hydrolysis direction remained almost constant over the entire pH range from 6.5 to 9, thus resembling the ATP hydrolysis pH profile. In contrast, proton transport through F_o in synthesis direction decreased continuously with increasing P-side pH values from 6 to 9, with half maximal velocity at pH 7.7 and less than 20% of the initial rate at pH 9. Thus, the pH profile for H^+ transport in synthesis direction resembles the ATP synthesis pH profile, except for a shift of the apparent pK_a from 7.7 to 6.2. This shift may be related to the demand of increased torque during ATP synthesis, where counteracting conformational restraints in the F_1 part have to be overcome. These restraints exposed by the F_1 part may also be responsible for the necessity of a minimum membrane potential of ~60 mV for ATP synthesis, but not for H^+ conduction through the isolated F_o complex.

Na^+ transport through F_o of the *P. modestum* ATP synthase has also been measured in both directions [54, 59]. In synthesis direction, Na^+ transport could be driven either by ΔpNa^+ or by $\Delta\psi$ and maximal rates were obtained, if the two driving forces were applied simultaneously. ATP synthesis and Na^+ transport through F_o in the appropriate direction thus responded to the same driving forces, except for the more stringent conditions for ATP synthesis, which required both driving forces simultaneously, whereas only one was sufficient for the ion conduction. In contrast, Na^+ transport through F_o in ATP hydrolysis direction was entirely dependent on $\Delta\psi$ and did not respond to the applied ΔpNa^+. Relating these results to the F_o motor model described above, they suggest that the arginine/binding site complex can only interact with ions entering from the periplasmic but not from the cytoplasmic side of the membrane. The arginine is therefore replaced from the complex with the binding site by high periplasmic but not by high cytoplasmic Na^+ concentrations. To break the complex from the cytoplasmic side obviously requires the membrane potential, which corroborates the notion that the potential interacts with the F_o motor through its cytoplasmic access route.

A New Model for the F_o Motor

In conclusion, these results suggest a novel mechanism for the F_o motor operation. In the resting state, an empty binding site near the periplasmic access channel of subunit a forms a complex with the stator arginine. At high periplasmic ion concentrations, the arginine can be pushed off the site by the coupling ion, causing dissociation between rotor and stator. The rotor has now freedom for rotation within a small angle, which may be sufficient to move the site with the newly bound ion into an area, where an exchange with ions in the cytoplasmic reservoir becomes feasible (not the cytoplasmic channel). After rotating backwards, the ion is striped off from the site and diffuses into the periplasmic channel, while the complex with the arginine reforms. Such an idling motion of the rotor versus the stator will lead to ion exchange between the two sides of the membrane, as observed experimentally for the Na^+ ATP synthase [59, 60].

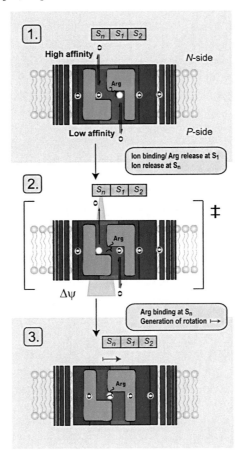

Figure 6. Events in the rotor/stator interface during ATP synthesis (rotation from left to right). Side view of the a/c interface of an F_0 motor. Shown is the c-ring (blue) with binding sites (red circles, S_1 to S_n) and subunit a containing the stator charge (Arg) and the (aqueous) P-side and the N-side access channels (green). In the starting position, the stator arginine is tightly bound to binding site S_1 **(1)**. Governed by chemical equilibria, which are influenced by the driving forces $\Delta\psi$ and $\Delta pH/\Delta pNa$, an incoming ion from the P-side is bound to site S_1 (requiring a high ion concentration), which displaces the Arg from the S_1 site. Simultaneously, the equilibria have to allow a release of an ion from the S_n site to the N-side, leading to a transition state in which the Arg is located between sites S_1 and S_n and the rotor is disconnected from the stator **(2)**. This allows the neutralized S_1 site to move into the lipid bilayer and the formation of an Arg-S_n complex in the interface, which generates unidirectional torque **(3)**. Browian motions are thought to bring the F_0 motor further into a new starting position (3. → 1.). See text for further details.

In the presence of appropriate driving forces, however, the motor will switch to unidirectional rotation and generate torque (Figure 6). In order to do so, the ion from the next incoming site must dissociate and escape through the cytoplasmic channel. This event is supported by a low ion concentration in the cytoplasmic reservoir, the positively charged arginine, and by the membrane potential. Attraction of the thus generated negatively charged, empty site by the positive stator charge causes rotation and generates torque. During the rotation step, the previously occupied site at the periplasmic channel is pushed out of the interface with subunit a and a new occupied site is pulled into the subunit a interface from the opposite side to contact the cytoplasmic access channel. The core feature of the new model is the so far unrecognized requirement for high coupling ion concentrations at the periplasmic side of the membrane. In the absence of these conditions, the motor is not functional, regardless of the magnitude of the applied driving forces. These new insights together with other observations suggest a torque generating mechanism, in which the electrostatic attraction of a negatively charged empty site by the positively charged arginine acts as the power stroke and where the driving forces serve to generate these positive and negative charges at the periplasmic and cytoplasmic channel, respectively.

CONCLUDING REMARKS

Finally, it should be emphasized that the F_o motor is highly asymmetric and appears to be perfectly constructed for $\Delta\mu H^+/\Delta\mu Na^+$-driven rotation in ATP synthesis direction. At sufficiently high periplasmic ion concentrations, the motor switches from its resting state with the arginine/binding site complex into a mobile state, where the binding site at the periplasmic channel has been occupied and the arginine has been liberated. This state provides a favourable provision for proceeding into ATP synthesis directed rotation, if the proper driving forces are applied. The only other option for the mobile state is to switch backwards into the resting state, if the periplasmic ion concentrations are reduced. For rotation into the opposite direction, it would be necessary to replace the arginine/binding site complex forming the resting state by ions from the cytoplasmic side of the membrane. Such a reaction can be excluded, however, at least for the Na^+ ATP synthase, since the rotation of the F_o motor in ATP hydrolysis direction was not driven by ΔpNa^+ [54, 59]. To prevent such unproductive movements into the wrong direction certainly increases the efficiency of the ATP synthase and may help to protect its structural integrity.

REFERENCES

[1]   Dimroth, P. (1997) Primary sodium ion translocating enzymes. *Biochim. Biophys. Acta* **1318**:11–51.
       doi: 10.1016/S0005-2728(96)00127-2.

[2]   Boyer, P.D. (1997) The ATP synthase – a splendid molecular machine. *Ann. Rev. Biochem.* **66**:717 – 749.
doi: 10.1146/annurev.biochem.66.1.717

[3]   Capaldi, R.A., Aggeler, R. (2002) Mechanism of the F_1F_o-type ATP synthase, a biological rotaty motor. *Trends Biochem. Sci.* **27**:154 – 160.
doi: 10.1016/S0968-0004(01)02051-5.

[4]   von Ballmoos, C., Cook, G.M., Dimroth, P. (2008) Unique rotary ATP synthase and its biological diversity. *Ann. Rev. Biophys.* **37**:43 – 64.
doi: 10.1146/annurev.biophys.37.032807.130018.

[5]   von Ballmoos, C., Wiedenmann, A., Dimroth, P. (2009) Essentials for ATP synthesis by F_1F_o ATP synthases. *Ann. Rev. Biochem.* **78**:649 – 672.
doi: 10.1146/annurev.biochem.78.081307.104803.

[6]   Abrahams, J.P., Leslie, A.G., Lutter, R., Walker, J.E. (1994) Structure at 2.8 Å resolution of F_1 ATPase from bovine heart mitochondria. *Nature* **370**:621 – 628.
doi: 10.1038/370621a0.

[7]   Boyer, P.D. (1993) The binding change mechanism for ATP synthase – some probabilities and possibilities. *Biochim. Biophys. Acta* **1140**:215 – 250.
doi: 10.1016/0005-2728(93)90063-L.

[8]   Noji, H., Yasuda, R., Yoshida, M., Kinosita Jr., K. (1997) Direct observation of the rotation of F_1-ATPase. *Nature* **386**:299 – 302.
doi: 10.1038/386299a0.

[9]   Yasuda, R., Noji, H., Yoshida, M., Kinosita Jr., K., Itoh, H. (2001) Resolution of distinct rotational substeps by submillisecond kinetic analysis of F_1-ATPase. *Nature* **410**:898 – 904.
doi: 10.1038/35073513.

[10] Adachi, K., Oiwa, K., Nishizaka, T., Furuike, S., Noji, H., Itoh, H., Yoshida, M., Kinosita Jr., K. (2007) Coupling of rotation and catalysis in F_1-ATPase revealed by single-molecule imaging and manipulation. *Cell* **130**:309 – 321.
doi: 10.1016/j.cell.2007.05.020.

[11] Kaim, G., Prummer, M., Sick, B., Zumofen, G., Renn, A., Wild, U.P., Dimroth. P. (2002) Coupled rotation within single F_oF_1 enzyme complexes during ATP synthesis and hydrolysis. *FEBS Lett.* **525**:156 – 163.
doi: 10.1016/S0014-5793(02)03097-1.

[12]   Diez, M., Zimmermann, B., Börsch, M., König, M., Schweinberger, E., Steigmiller, S., Reuter, R., Felekyan, S., Kudryavtsev, V., Seidel, C.A., Gräber, P. (2004) Proton-powered subunit rotation in single membrane-bound F_oF_1 ATP synthase. *Nat. Struct. Mol. Biol.* **11**:135 – 141.
       doi: 10.1038/nsmb718.

[13]   Düser, M.G., Zarrabi, N., Cipriano, D.J., Ernst, S., Glick, G.D., Dunn, S.D., Börsch, M. (2009) 36° step size of proton-driven c-ring rotation in F_oF_1-ATP synthase. *EMBO J.* **28**:2689 – 2696.
       doi: 10.1038/emboj.2009.213.

[14]   Stock, D., Leslie, A.G., Walker, J.E. (1999) Molecular architecture of the rotary motor in ATP synthase. *Science* **286**:1700 – 1705.
       doi: 10.1126/science.286.5445.1700.

[15]   Pogoryelov, D., Nikolaev, Y., Schlatter, U., Pervushin, K., Dimroth, P., Meier, T. (2008) Probing the rotor subunit interface of the ATP synthase from *Ilyobacter tartaricus*. *FEBS J.* **275**:4850 – 4862.
       doi: 10.1111/j.1742-4658.2008.06623.x.

[16]   Valiyaveetil, F.I., Fillingame, R.H. (1998) Transmembrane topography of subunit a in the *Escherichia coli* F_1F_o ATP synthase. *J. Biol. Chem.* **273**:16241 – 16247.
       doi: 10.1074/jbc.273.26.16241.

[17]   Vik, S.B., Ishmukhametov, R.R. (2005) Structure and function of subunit a of the ATP synthase of *Escherichia coli*. *J. Bioenerg. Biomembr.* **37**:445 – 449.
       doi: 10.1007/s10863-005-9488-6.

[18]   Dunn, S., Revington, M., Cipriano, D., Shilton, B. (2000) The b subunit of *Escherichia coli* ATP synthase. *J. Bioenerg. Biomembr.* **32**:347 – 355.
       doi: 10.1023/A:1005571818730.

[19]   Greie, J.C., Deckers-Hebestreit, G., Altendorf, K. (2000) Subunit organization of the stator part of the F_o complex from *Escherichia coli* ATP synthase. *J. Bioenerg. Biomembr.* **32**:357 – 364.
       doi: 10.1023/A:1005523902800.

[20]   Weber, J. (2007) ATP synthase – the structure of the stator stalk. *Trends Biochem. Sci.* **32**:53 – 56.
       doi: 10.1016/j.tibs.2006.12.006.

[21]   Dickson, V.K., Silvester, J.A., Fearnley, I.M., Leslie, A.G., Walker, J.E. (2006) On the structure of the stator of the mitochondrial ATP synthase. *EMBO J.* **25**:2911 – 2918.
       doi: 10.1038/sj.emboj.7601177.

[22] Meier, T., Polzer, P., Diederichs, K., Welte, W., Dimroth, P. (2005) Structure of the rotor ring of F-type Na⁺-ATPase from *Ilyobacter tartaricus*. *Science* **308**: 659 – 662. doi: 10.1126/science.1111199.

[23] Oberfeld, B., Brunner, J., Dimroth, P. (2006) Phospholipids occupy the internal lumen of the c ring of the ATP synthase of *Escherichia coli*. *Biochemistry* **45**:1841 – 1851. doi: 10.1021/bi052304+.

[24] Vonck, J., Krug von Nidda, T., Meier, T., Matthey, U., Mills, D.J., Kühlbrandt, W., Dimroth, P. (2002) Molecular architecture of the undecameric rotor of a bacterial Na⁺-ATP synthase. *J. Mol. Biol.* **321**:307 – 316. doi: 10.1016/S0022-2836(02)00597-1.

[25] von Ballmoos, C., Appoldt, Y., Brunner, J., Granier, T., Vasella, A., Dimroth, P. (2002) Membrane topography of the coupling ion binding site in Na⁺-translocating F₁F₀ ATP synthase. *J. Biol. Chem.* **277**:3504 – 3510. doi: 10.1074/jbc.M110301200.

[26] Meier, T., Krah, A., Bond, P.J., Pogoryelov, D., Diederichs, K., Fernaldo-Gomez, J.D. (2009) Complete coordination structure in the rotor ring of Na⁺-dependent F-ATP synthase. *J. Mol. Biol.* **391**:498 – 507. doi: 10.1016/j.jmb.2009.05.082.

[27] Murata, T., Yamato, I., Kakinuma, Y., Leslie, A.G., Walker, J.E. (2005) Structure of the rotor of the V-type Na⁺-ATPase from *Enterococcus hirae*. *Science* **308**:654 – 659. doi: 10.1126/science.1110064.

[28] Pogoryelov, D., Yildiz, O., Fernaldo-Gomez, J.D., Meier, T. (2009) High-resolution structure of the rotor ring of a proton-dependent F-ATP synthase. *Nat. Struct. Mol. Biol.* **16**:1068 – 1073. doi: 10.1038/nsmb.1678.

[29] Preiss, L., Yldiz, O., Hicks, D.B., Krulwich, T.A., Meier, T. (2010) A new type of proton coordination in an F₁F₀-ATP synthase rotor ring. *PLoS Biology* **8**:1 – 10. doi: 10.1371/journal.pbio.1000443.

[30] Dautant, A., Velours, J., Giraud, M.F. (2010) Crystal structure of the Mg.ADP-inhibited state of the yeast $F_1 c_{10}$ ATP synthase. *J. Biol. Chem.* **285**(38):29502 – 29510. doi: 10.1074/jbc.M110.124529.

[31] von Ballmoos, C., Dimroth, P. (2007) Two distinct proton binding sites in the ATP synthase family. *Biochemistry* **46**:11800 – 11809. doi: 10.1021/bi701083v.

[32]   Krah, A., Pogoryelov, D., Langer, J., Bond, P.J., Meier, T., Feraldo-Gomez, J.D. (2010) Structural and energetic basis for H^+ versus Na^+ binding selectivity in ATP synthase F_o rotors. *Biochim. Biophys. Acta* **1797**:763–772. doi: 10.1016/j.bbabio.2010.04.014.

[33]   Meier, T., Matthey, U., von Ballmoos, C., Vonck, J., Krug von Nidda, T., Kühlbrandt, W., Dimroth, P. (2003) Evidence for structural integrity in the undecameric c-rings isolated from sodium ATP synthases. *J. Mol. Biol.* **325**:389–397. doi: 10.1016/S0022-2836(02)01204-4.

[34]   von Ballmoos, C., Meier, T., Dimroth, P. (2002) Membrane-embedded location of Na^+ or H^+ binding sites on the rotor ring of F_1F_o ATP synthases. *Eur. J. Biochem.* **269**:5581–5589. doi: 10.1046/j.1432-1033.2002.03264.x.

[35]   Junge, W., Lill, H., Engelbrecht, S. (1997) ATP synthase: an electrochemical transducer with rotary mechanics. *Trends Biochem. Sci.* **22**:420–423. doi: 10.1016/S0968-0004(97)01129-8.

[36]   Angevine, C.M., Herold, K.A., Fillingame, R.H. (2003) Aqueous access pathways in subunit a of rotary ATP synthase extend to both sides of the membrane. *Proc. Natl. Acad. Sci. U.S.A.* **100**:131779–13183. doi: 10.1073/pnas.2234364100.

[37]   Angevine, C.M., Fillingame, R.H. (2003) Aqueous access channels in subunit a of rotary ATP synthase. *J. Biol. Chem.* **278**:6066–6074. doi: 10.1074/jbc.M210199200.

[38]   Steed, P.R., Fillingame, R.H. (2009) Aqueous accessibility of the transmembrane regions of subunit c of the *Escherichia coli* F_1F_o ATP synthase. *J. Biol. Chem.* **284**:23243–23250. doi: 10.1074/jbc.M109.002501.

[39]   Jiang, W., Fillingame, R.H. (1998) Interacting helical faces of subunits a and c in the F_1F_o ATP synthase of *Escherichia coli* defined by disulfide cross-linking. *Proc. Natl. Acad. Sci. U.S.A.* **95**:6607–6612. doi: 10.1073/pnas.95.12.6607.

[40]   Vorburger, T., Zingg Ebneter, J., Wiedenmann, A., Morger, D., Weber, G., Diederichs, K., Dimroth, P., von Ballmoos, C. (2008) Arginine-induced conformational change in the c-ring/a-subunit interface of the ATP synthase. *FEBS J.* **275**:2137–2150. doi: 10.1111/j.1742-4658.2008.06368.x.

[41] Lau, W.C.Y., Rubinstein, J.L. (2010) Structure of intact *Thermus thermophilus* V-ATPase by cryo-EM reveals organization of the membrane-bound V_o motor. *Proc. Natl. Acad. Sci. U.S.A.* **107**:1367–1372.
doi: 10.1073/pnas.0911085107.

[42] Cain, B.D., Simoni, R.D. (1989) Proton translocation by the F_1F_o ATPase of *Escherichia coli*. Mutagenic analysis of the a subunit. *J. Biol. Chem.* **264**:3292–3300.

[43] Wehrle, F., Kaim, G., Dimroth, P. (2002) Molecular mecuanism of the ATP synthase's F_o motor probed by mutational analyses of subunit a. *J. Mol. Biol.* **322**:369–381.
doi: 10.1016/S0022-2836(02)00731-3.

[44] Xing, J., Wang, H., von Ballmoos, C., Dimroth, P., Oster, G. (2004) Torque generation by the F_o motor of the sodium ATPase. *Biophys. J.* **87**:2148–2163.
doi: 10.1529/biophysj.104.042093.

[45] Mitchell, P. (1961) Coupling of phosphorylation to electron and hydrogen transfer by a chemi-osmotic type of mechanism. *Nature* **191**:144–148.
doi: 10.1038/191144a0.

[46] Jagendorf, A.T., Uribe, E. (1966) ATP formation caused by acid-base transition of spinach chloroplasts. *Proc. Natl. Acad. Sci. U.S.A.* **55**:170–177.
doi: 10.1073/pnas.55.1.170.

[47] Kaim, G., Dimroth, P. (1999) ATP synthesis by F-type ATP synthase is obligatorily dependent on the transmembrane voltage. *EMBO J.* **18**:4118–4127.
doi: 10.1093/emboj/18.15.4118.

[48] Mitchell, P. (1966) Chemiosmotic coupling in oxidative and photosynthetic phosphorylation. *Biol. Rev. Cam. Philos. Soc.* **41**:445–502.
doi: 10.1111/j.1469-185X.1966.tb01501.x.

[49] Cockrell, R.S., Harris, E.J., Pressman, B.C. (1967) Synthesis of ATP driven by a potassium gradient in mitochondria. *Nature* **215**:1487–1488.
doi: 10.1038/2151487a0.

[50] Maloney, P.C., Wilson, T.H. (1975) ATP synthesis driven by a proton motive force in *Streptococcus lactis*. *J. Membr. Biol.* **25**:285–310.
doi: 10.1007/BF01868580.

[51] Sone, N., Yoshida, M., Hirata, H., Kagawa, Y. (1977) Adenosine triphosphate synthesis by electrochemical proton gradient in vesicles reconstituted from purified adenosine triphosphatase and phospholipids from thermophilic bacterium. *J. Biol. Chem.* **252**:2956–2960.

[52]   Fischer, S., Gräber, P. (1999) Comparison of ΔpH-and Δψ-driven ATP synthesis catalyzed by the ATPase from *Escherichia coli* or chloroplasts reconstituted into liposomes. *FEBS Lett.* **457**:327 – 332.
doi: 10.1016/S0014-5793(99)01060-1.

[53]   Etzold, C., Deckers-Hebestreit, G., Altendorf, K. (1997) Turnover number of *Escherichia coli* F_oF_1 ATP synthase in membrane vesicles. *Eur. J. Biochem.* **243**:336 – 343.
doi: 10.1111/j.1432-1033.1997.0336a.x.

[54]   Wiedenmann, A., Dimroth, P., von Ballmoos, C. (2009) Functional asymmetry of the F_o motor in bacterial ATP synthases. *Mol. Microbiol.* **72**:479 – 490.
doi: 10.1111/j.1365-2958.2009.06658.x.

[55]   Williams, R.J. (1978) The multifarious coupling of energy transduction. *Biochim. Biophys. Acta* **505**:1 – 44.

[56]   Laubinger, W., Dimroth, P. (1988) Characterization of the ATP synthase of *Propionigenium modestum* as a primary sodium pump. *Biochemistry* **27**:7531 – 7537.
doi: 10.1021/bi00419a053.

[57]   Kluge, C., Dimroth, P. (1993) Kinetics of inactivation of the F_1F_o ATPase of *Propionigenium modestum* by dicyclohexylcarbodiimide in relationship to H^+ and Na^+ coordination: probing the binding site for thje coupling ions. *Biochemistry* **32**:10378 – 10386.
doi: 10.1021/bi00090a013.

[58]   Wiedenmann, A., Dimroth, P., von Ballmoos, C. (2008) Δψ and ΔpH are equivalent driving forces for proton transport through isolated F_o complexes of ATP synthases. *Biochim. Biophys. Acta* **1777**:1301 – 1310.
doi: 10.1016/j.bbabio.2008.06.008.

[59]   Kluge, C., Dimroth, P. (1992) Studies on Na^+ and H^+ translocation through the F_o part of the Na^+-translocating F_1F_o ATPase from *Propionigenium modestum*: discovery of a membrane potential dependent step. *Biochemistry* **31**:12665 – 12672.
doi: 10.1021/bi00165a017.

[60]   Kaim, G., Dimroth, P. (1998) Voltage-generated torque drives the motor of the ATP synthase. *EMBO J.* **17**:5887 – 5895.
doi: 10.1093/emboj/17.20.5887.

Reflections on Energy Conversion in Biological and Biomimetic Systems

Athel Cornish-Bowden[*], María Luz Cárdenas and Élisabeth Lojou

Unité de Bioénergetique et Ingénierie des Protéines,
Centre National de la Recherche Scientifique,
31 chemin Joseph-Aiguier, B.P.71, 13402 Marseilles, France

E-Mail: *acornish@ibsm.cnrs-mrs.fr

Received: 3rd September 2010 / Published: 13th June 2011

Abstract

In principle any form of energy (light, electrical, potential, chemical, kinetic energy, etc.) can be converted into any other, and a large part of biochemistry is concerned with the mechanisms of transduction. Despite this, misleading statements such as "glucose phosphorylation is coupled to ATP hydrolysis" appear even in modern books that appear in general to be based on a thorough understanding of thermodynamics. In reality, harnessing the chemical energy contained in an ATP molecule to drive metabolism involves no hydrolysis at all, and it is exactly because there is no hydrolysis that the process can work. At a grosser level, many authors still write as if production of mechanical work in organisms – from the packaging motor of bacteriophage to the muscles of large animals – operated in much the same way as industrial motors, i. e. that they release chemical energy as heat, which is then converted into work by the sort of pressure-volume effects discussed in elementary thermodynamics courses, but living motors – including not only muscles but also such examples as the DNA packaging motor of bacteriophage Φ29 – are not heat engines. ATP hydrolysis is, of course, a net effect but the heat that is produced is lost: it cannot be converted into work because organisms have no known mechanisms for transforming heat into pressure-volume work, at least, not on a significant scale. Unfortunately, elementary courses tend to concentrate on the thermodynamics of gases to such an extent that the irrelevance of pressure-volume work to biochemistry is completely lost, and the Gibbs energy, for example, is seen as having something to do with heat, even though its main role in isothermal systems is as a device for expressing equilibrium constants on a logarithmic scale. Thorough understanding of these concepts will be essential for the successful development of new biotechnologies. The case of biohydrogen as a fuel is discussed.

INTRODUCTION

All biological processes depend on the management of energy, and all energy except, in recent years, energy from nuclear power plants, arrives to the earth in the form of sunlight. Photosynthetic organisms convert sunlight directly into chemical energy, and some organisms convert chemical energy directly into light. Animals and other motile organisms transform chemical energy into movement, and kinetic and potential energy are easily interconvertible with negligible losses.

Most industrial processes, however, remain primitive by comparison. Transformations between potential energy, kinetic energy and electrical energy can be made directly, and with high efficiency, but other transformations are either very inefficient or possible only on a small scale, or both. Energy harnessing on an industrial scale normally involves heat production by combustion of fuels, followed by conversion of the heat into work, chemical energy or electrical power, and the inevitable energy losses are large. A heat engine working as a reversible Carnot cycle over the widest temperature range practical for a modern car, say from about 25 °C (298 K) up to about 120 °C (393 K) would convert about 24% of the heat consumed into work. In reality, of course, a real engine cannot operate under reversible conditions, and the practical efficiency of the best heat engines is more like 10%.

By comparison, a running animal such as a cheetah can convert the chemical energy of ATP into kinetic energy with about 50% efficiency. Of course, the ATP needs to be produced from food and oxygen by metabolic processes that involve additional losses, so the overall efficiency of conversion of food and oxygen into kinetic energy is around 20%. That is no less true of the estimation of the efficiency of a heat engine, for which the fuel likewise needs to be mined and refined before it can be used, so the estimate of a maximum efficiency of 10% is also over-optimistic.

These considerations have led to great and increasing interest in the possibility of mimicking the efficiency of green plants for converting sunlight into chemical energy, and the efficiency of animals for converting chemical energy into motion. The thermodynamic inefficiency of heat engines is, moreover, just one of the serious problems with current industrial processes. The most readily available fuels today are hydrocarbons, and problems of supply are already evident, for example, in the need to drill for oil in ever deeper water, and will only get worse. In addition, combustion of hydrocarbons produces not only water, which is largely harmless, but also CO_2, the cause of increasingly severe effects on the temperature of the atmosphere. So we need not only processes of greater efficiency, but also ones that do not produce any more CO_2 than they consume.

COMMON ERRORS IN THERMODYNAMIC ANALYSIS

Development of alternative fuels, such as H_2, and processing them with high energy efficiency, requires a proper understanding of some central thermodynamic concepts, which may appear elementary but are often forgotten. In this section, therefore, we discuss some problems that occur often in the literature of biochemistry, and make it difficult to make a proper appreciation of the thermodynamic constraints that need to be understood for improving the efficiency of processes for harnessing energy.

Measurement of entropies of reaction or of activation

Textbooks, even ones that are in many respects among the best available, commonly present the determination of an entropy of reaction from a van 't Hoff plot (the logarithm of an equilibrium constant plotted as a function of the reciprocal absolute temperature) as something easy and straightforward to do, because at superficial examination the required entropy value follows easily from the intercept on the ordinate axis. The problem, however, is that the required intercept is typically far away from the range of experimental observations, so a very long extrapolation is needed, as pointed out many years ago [2]. The difficulty does not go away if one uses computer fitting rather than a graph: it just becomes less obvious that there is a problem, and hence easier to deceive oneself. It is not a special problem of the van 't Hoff plot: essentially the same applies to the determination of an entropy of activation from an Arrhenius plot of the logarithm of a rate constant against the reciprocal absolute temperature [3 – 5].

To take a specific example, Figure 4.11 of [6] illustrates a van 't Hoff plot with an unlabelled scale of reciprocal temperature, but one that implicitly starts at zero, because the intercept on the vertical axis is labelled as $\Delta S^0/R$, in which case the observations span an impressive 15-fold range of absolute temperature. However, that is obviously impossible, at least for a biological process: if we assume that the higher temperature was about 85 °C (358 K), a typical temperature for a thermophilic organism, the lower temperature must have been – 250 °C (24 K); if, on the other hand, the lower temperature was – 10 °C (263 K), a typical temperature for a psychrophilic organism, the higher must have been 3700 °C (4000 K). Neither of these interpretations is plausible, so we must suppose that the figure did not represent a real or feasible experiment, but a purely imaginary one.

If the ordinate axis in a van 't Hoff or Arrhenius plot is drawn at the true zero on the $1/T$ axis it becomes evident that estimation of the intercept in any real experiment implies an extrapolation that is typically of the order of $10-15$ times the range of the observations, and therefore cannot be done accurately. For example, in a study of the ribosome featured on the cover of the *Proceedings of the National Academy of Sciences* [7] the $T\Delta S^{\ddagger}$ for peptide bond formation was reported to be accurate to $\pm 10\%$ but as this required an extrapolation of 13 times the range of observations in an Arrhenius plot, it appears unduly optimistic.[1]

The only reasonable conclusions to be drawn are as follows:

1. The entropy of reaction cannot be obtained with usable precision from a van 't Hoff plot;

2. The entropy of activation cannot be obtained with usable precision from an Arrhenius plot.

If these parameters are needed they must be determined in some other way, such as from precise calorimetric measurements.

Metabolic "efficiency"

The notion of metabolic efficiency is less frequently encountered in modern textbooks and papers in biochemistry than it once was, but it is still found more frequently than it ought to be. Statements such as

Dividing the $\Delta G^{0\prime}$ of ATP formation by that of lactate formation indicates that homolactic fermentation is 31% "efficient", that is, 31% of the free energy released by this process under standard biochemical conditions is sequestered in the form of ATP. [8]

have by no means disappeared from the literature. A related but perhaps less objectionable sort of statement that remains common is

Glucose phosphorylation is coupled to ATP hydrolysis. [6]

The problem with this last statement is that the process concerned is the reaction between glucose and ATP to produce glucose 6-phosphate and ADP, which does not involve hydrolysis: no water participates in the overall reaction (contrast this with the ATP hydrolysis used to drive the DNA packaging motor described in *Energy Transduction in Living Organisms*, see below). This may be justified as useful shorthand for calculating Gibbs energies of

---

[1] By chance this paper was published the same week as a previous Beilstein Symposium at which this topic was being discussed [5].

reaction: the net Gibbs energy for glucose phosphorylation can certainly be calculated as the difference between the Gibbs energies of hydrolysis of glucose and of ATP, and as long as it is limited to that no harm is done. The trouble comes when a legitimate difference in Gibbs energies is taken to sanction a completely illegitimate ratio of Gibbs energies. Atkinson [9] pointed out long ago that a statement such as the first one quoted is "false in every part", but his warning has yet to be generally heeded.

The confusion arises from two sources, the introduction of an irrelevant species, water, and the arbitrary definition of the standard state as 1 mol/l. The calculation of "efficiency" can be done just as easily with a different irrelevant species, such as fructose, or with a different choice of standard state for removing the units from the calculation of the logarithms of concentrations, and in either of these cases the calculated "efficiency" of a process is quite different. With fructose instead of water as the irrelevant species, for example, the "efficiency" of the hexokinase reaction is changed from 45% to − 14%; taking the standard state of all species as 0.1 μmol/l instead of 1 mol/l it would be 76% [10]. In effect, with suitable choices of irrelevant species and standard states one can obtain any "efficiency" one wishes.[2]

It is important to avoid this sort of confusion, because consideration of thermodynamic efficiency has been an essential part of the design of machines since the time of Carnot, and will certainly be essential for developing better ways of harnessing energy in the future, but it needs to be done properly.

ENERGY TRANSDUCTION IN LIVING ORGANISMS

As we have noted, animals can convert chemical energy into kinetic or potential energy with efficiency at least as high as any machine. This capacity is not confined to the animal kingdom, and a striking example of an extremely small motor is provided by the DNA packaging motor of the bacteriophage $\Phi29$, which converts the energy stored in ATP molecules directly into pressure-volume work, achieving a pressure of the order of 50 atmospheres (about ten times the pressure in a bottle of champagne). The way in which this is achieved has been studied in detail in single-molecule studies [11]. It involves a high degree of coordination between the subunits of the ring ATPase present in the bacteriophage. Homomeric ring ATPases are found in all forms of life and are involved in many processes such as chromosome segregation, protein unfolding and ATP synthesis. In the bacteriophage it is used to load the double-stranded DNA genome into the viral shell. A high precision

---

[2] One of us presented these ideas at a meeting in 1982 [10], after which the distinguished expert in network thermodynamics Jörg Stucki commented that although he thought they were correct, he also doubted whether anyone would make such absurd calculations. A year later, at another meeting, he reported that he had checked the literature and had found that such calculations are, in fact, quite common.

single-molecule assay involving dual-beam "optical tweezers", have shown that the five-member motor packages the DNA in 10-base-pair bursts, each consisting of four individual 2.5-bp steps corresponding to the hydrolysis of a single ATP. They thus provide a direct measurement of a single enzymatic cycle by this ATPase, revealing an unexpected form of coordination between the subunits. The non-integral step size detected is unprecedented, and raises intriguing mechanistic questions about ATP hydrolysis within rings, and the interactions of the ring with DNA.

Note that in this engine the DNA packaging is driven by ATP hydrolysis, in contrast to the sort of reactions discussed in *Metabolic "efficiency"* (see above), in which water is not a substrate. Nonetheless, it still contrasts with the heat engines used in industry, because the heat produced is simply lost, and is not used to generate pressure-volume work. Although the same principles of thermodynamics apply both to the steam engines studied by Carnot and to energy transduction in organisms, the analogy must not be taken too far, because heat engines do not exist (as far as we know) in organisms.

PHOTODECOMPOSITION OF WATER

The photodecomposition of water by photosynthesis has several characteristics that make it very attractive for the future development of electrical power: it depends only on sunlight as energy source, which is available in vast amounts, far more than any realistic requirement for human activities; it does not produce CO_2 but instead consumes it; it is thermodynamically efficient, as nearly all the light of the appropriate wavelength falling on a leaf is absorbed, and about 50% of the light absorbed is converted into chemical energy [1]. On the other hand, it is far more complicated than a typical industrial process, involving many steps and many catalysts. The subsequent combustion of the glucose produced is likewise complicated, with numerous steps and catalysts. Glucose is thermodynamically very unstable in the presence of O_2, but it is kinetically very stable. In solid form it can survive being kept in dry air for years. In solution it is slowly converted to CO_2 and water, but because of bacterial contamination, not through any direct effect of the water and O_2. In brief, oxidation of glucose can only proceed in mild conditions in the presence of a catalyst.

The need for many steps – each with its own catalyst – is a consequence of chemical constraints in the form of a need to pass through chemically feasible intermediate metabolites, and the need to extract the energy in small packages. Analysis of other metabolic pathways [12] indicates that although pathways shorter than those that exist in living organisms may be feasible, they typically suffer from the need to pass through thermodynamically unfavourable steps.

At the level of small-scale research, photocells are beginning to approach the performance of photosynthesis, the best multijunction concentrators now being reported by the U.S. National Renewable Energy Laboratory [13] to have a thermodynamic efficiency greater than

40%. This is very encouraging, but photocells cannot solve all transportation needs without passing through a chemical step: aeroplanes cannot convert electrical power directly into motion, because they need a transportable energy source. Private cars of the future may perhaps be able to obtain electrical power directly from an external supply as present-day trains do, but at present they require rechargeable chemical stores, whether batteries or fuel cells. In any case, it is hard to imagine that aeroplanes will ever escape the need for chemical stores.

H$_2$ AS FUEL

General considerations

Great efforts are being devoted at present to find ways of escaping from our current reliance on fossil fuels and the production of CO_2, and to develop new sources of power. One of these is H_2, which has several advantages over hydrocarbons as a potential fuel:

1. It can in principle be produced by the photodecomposition of water driven by sunlight, available in unlimited quantities.

2. It can also be produced by electrolysis of water, ideally *in situ* at a hydroelectric power plant: this primary source of energy is not unlimited, but it is renewable.

3. The raw material, water, is abundantly available.

4. Neither the production nor the use of H_2 implies production of CO_2.

However, it also has some associated problems:

1. Much more efficient processes than those available today will be required, not only for production of H_2 by photodecomposition or electrolysis of water, but also for conversion of H_2 into work, in fuel cells. (Heat engines are already available, of course, but no major improvements in thermodynamic efficiency can be expected.)

2. H_2 is much more difficult to store safely than hydrocarbons, which are themselves responsible for many accidents.

3. H_2 leaks through pores in containers that can be considered perfectly hermetic for other fluids.

4. Efficient production and harnessing of H_2 depends on catalysts.

Some of these problems will certainly be solved in future years, and we can expect that the current dependence on heat engines and CO_2 production will be overcome with cyclic processes in which energy from sunlight, wind and hydrostatic potential is used to generate

H_2 that is subsequently oxidized to water with the generation of work. Before this can be achieved, however, a major obstacle will need to be removed: catalysts that can be used on an industrial scale will be required.

Fuel cells

In fuel cells based on H_2 the fuel is oxidized at the anode side, and the electrons that are released by the oxidation reaction are driven through an outer electrical circuit, thus generating electrical current. The electrons reach the cathode, where they combine with an oxidant (typically O_2) and protons to a product (typically water). Enzymatic fuel cells employ enzymes instead of noble metals to catalyse the reaction. Fuel cells or biofuel cells thus present similar problems to those of photodecomposition or electrolysis of water. Use of biocatalysts in fuel cells provides, however, several advantages over catalysis by noble metals. Biocatalysts are abundantly available, in contrast to the limited availability of transition-metal catalysts. The substrate specificity diminishes reactant cross-over, which in theory would allow membraneless fuel cells, and biofuel cells can function at neutral pH and moderate temperatures. Catalysts for the oxygenation are available in the form of the enzyme bilirubin oxidase, but the first step is much more problematic, because although hydrogenases are available from H_2-producing organisms, the most abundant ones are sensitive to O_2.

H_2-producing bacteria occur mainly in anaerobic environments. For example, *Desulfovibrio fructosovorans*, a mesophilic sulphate-reducing bacterium, contains both Fe and NiFe enzymes, both of which are extremely sensitive to O_2 [14]. Nonetheless, there are also some microorganisms found in aerobic environments, such as the pollution tolerant bacteria *Ralstonia metallidurans* and *Ralstonia eutropha* [15], *Escherichia coli* [16] and the hyperthermophilic bacterium *Aquifex aeolicus* [17]. This last grows at 85 °C, and is thus particularly attractive for potential industrial use.

Catalysts

Useful catalysts for production of H_2 driven by sunlight etc., or for conversion of H_2 into electrical power, both of these on an industrial scale, must satisfy three criteria:

1. It must be available in quantity at a modest price: Pt, currently costing almost 50 Euros per gram, is clearly excluded.

2. It must be robust enough to withstand long-term use in industrial conditions: this eliminates many enzymes. Resistance to O_2 may not be absolutely essential, but it is certainly highly desirable.

3. It must be energy-efficient, with negligible losses due to hysteresis. Pt is very efficient in this sense, whereas more readily available metals, such as Co and Ni, are much less so.

4. It must be efficiently immobilized at the electrode surface. This is often a bottleneck, especially for enzymes that need specific electrode modification for establishing electrical contact (in other words, nature did not evolve enzymes for bioelectroanalytical applications).

Unfortunately the best metallic catalysts, Pt as a general catalyst, and Pd as a catalyst especially adapted to reactions of H_2, are among the six or so least abundant elements in the earth's crust, each with an abundance of about 10^{-10} atoms per atom of Si [18]. They clearly cannot form the basis of a large-scale industrial process, either now or in the future. Lighter metals from the same groups of the periodic table, such Co and Ni, catalyse similar reactions, but much less efficiently and with much greater energy losses.

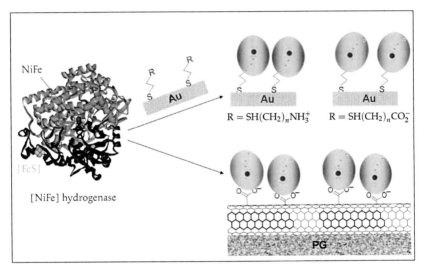

Figure 1. Orientation of *Desulfovibrio fructosovorans* NiFe hydrogenase at self-assembled monolayer gold electrodes or pyrolytic graphite (PG) electrode modified by a carbon nanotube network. Due to the dipole moment of the enzyme and the acidic environment of the surface FeS cluster, switching of the enzyme is achieved by varying the charge of the interface. The electron transfer process consequently shifts from a direct to a mediated electrical communication.

On the other hand they are far more abundant, at more than 10^{-5} atoms per atom of Si, and far less expensive, in the range 15 – 30 Euros per kg. The question therefore arises as to whether biomimetic catalysts, in which metal ions are complexed with organic ligands designed to resemble the active sites of suitable enzymes, might combine the energy efficiency of Pt and some enzymes with a much greater stability than is possible with natural enzymes. Fontecave and collaborators have made substantial progress towards this goal with

Ni-based carbon nanotube coated electrodes that exhibit high catalytic activity for H_2 evolution in aqueous solutions, and also high current densities for H_2 oxidation, though at large overpotentials [19]. This is a promising approach, and even if biomimetic catalysts are not ready to be used on an industrial scale today there is a realistic possibility that they will become so.

Immobilization of hydrogenase

A different approach is to improve hydrogenase immobilization onto structurally modified electrodes. In recent years remarkable successes have been obtained in different fields that contribute to the efficient electrical communication between the enzyme and the electrode. These successes relate to enzyme stability, co-immobilization of the enzyme with a redox mediator, control of the enzyme orientation and increase in the enzyme density, as now outlined:

1. Enzyme stability has been improved by immobilizing a membrane-bound hyperthermophilic hydrogenase inserted into liposomes. A five-fold increase in the enzymatic efficiency is observed, mainly due to the reconstitution of a physiological like environment [20].

2. Co-entrapment of a NiFe mesophilic hydrogenase with a redox mediator (either methyl viologen or the hydrogenase physiological partner) in clay films deposited at the surface of an electrode shows satisfactory catalytic efficiencies both for H_2 evolution and uptake [21].

3. The modification of the electrode so that a redox relay at the protein periphery is located at a tunnel distance from the electrode surface allows efficient H_2 oxidation with no need of a redox mediator (Fig. 1). This situation is encountered through chemical modification of self-assembled-monolayer gold electrodes to match the enzyme surface close to the redox relay [22]. A switching in space has been demonstrated for the O_2-sensitive NiFe hydrogenase from *Desulfovibrio fructosovorans*, which has a high dipole moment and a group of negatively charged residues around a FeS cluster located at the surface of the enzyme.

4. The size of the hydrogenase results in a low density of catalytic centres at a smooth electrode surface. The development of 3D architectures provides new basis for efficient H_2 oxidation. Both long-term stability and catalytic efficiency are improved by modification of the electrode by carbon nanotube deposits. Further physical [22] or chemical [23] modification of the nanotubes has been found to allow the control of the orientation of the hydrogenase on each nanotube, and thus the electron transfer process. The association of carbon nanotube films developing large surface areas and numerous anchoring sites with the capability of the hyperthermophilic hydrogenase from *Aquifex aeolicus* to oxidize H_2 at elevated temperatures (up to 85°C) yields current densities as illustrated in

Figure 2. Values up to 1 mA/cm^2 have been obtained. Encapsulation of the hydrogenase in a carbon nanotube network prior to immobilization on the electrode surface permits stabilization of the catalytic signal with time. H$_2$ oxidation occurs even in the presence of O$_2$ and CO [24]. Work is now in progress towards lowering the cost of hydrogenase production, and using a matrix other than toxic carbon nanotubes.

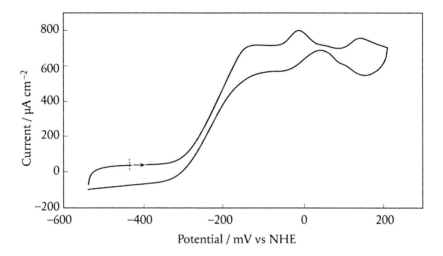

Figure 2. Cyclic voltammogram recorded at a pyrolytic graphite electrode modified with an amino-functionalized carbon nanotube film, under H$_2$ at 60 °C. The voltammogram shows the catalytic oxidation of H$_2$ by the hyperthermophilic hydrogenase from *Aquifex aeolicus* encapsulated in the carbon nanotube network. The dashed vertical line indicates the potential of the H$^+$/H^2 couple under the experimental conditions (HEPES buffer, pH 7.2, 10 mV s^{-1}), and the arrow indicates the direction of the scan.

PERSPECTIVES

It will be clear from these examples that a H$_2$-based economy is not yet ready to replace the existing dependence on hydrocarbons. The use of chemical processes that 9 mimic enzyme action is promising, but the performance in aqueous solution needs to be improved. On the other hand, isolated enzymes present problems of stability and O$_2$ sensitivity, though attempts are in progress to overcome this difficulty [25]. A third possibility is to harness the H$_2$-generating capacity of living organisms, but it should never be forgotten that all of these organisms have been endowed by billions of years of evolution with regulatory mechanisms that work for their own good, not for the good of humanity, so the realistic potential of this approach is far from being established. Many attempts to use microorganisms in biotechnological processes have failed completely, or have at best been very disappointing, because the regulatory mechanisms have proved to be very effective at resisting efforts to force metabolic flux towards commercially valuable metabolites that are not valuable for the microbe itself. Suppressing the regulation is possible, but it results in an unhealthy microbe, so although it may work to some degree [26] in the short-term it cannot be the basis for an industrial process, and the biomimetic strategy seems more promising.

At present we are studying the possibility of using consortia of microbes, in which the different partners produce metabolites needed by the others, and the overall effect is to generate H_2 from cheap raw materials, such as domestic waste [27].

REFERENCES

[1]   Dau, H., Zaharieva, I. (2009) Principles, efficiency, and blueprint character of solar-energy conversion in photosynthetic water oxidation. *Acc. Chem. Res.* **42**:1861 – 1870.
doi: 10.1021/ar900225y

[2]   Exner, O. (1964) On the enthalpy – entropy relationship. *Coll. Czech. Chem. Comm.* **26**:1094 – 1113.

[3]   Cornish-Bowden, A. (2002) Enthalpy – entropy compensation: a phantom phenomenon. *J. Biosci.* **27**:121 – 126.
doi: 10.1007/BF02703768

[4]   Cornish-Bowden, A. (2004) Fundamentals of Enzyme Kinetics (3rd edn.), pp. 235 – 237, Portland Press, London.

[5]   Cornish-Bowden, A., Cárdenas, M.L. (2004) Handling Equilibrium Processes Embedded in Metabolic Systems. In: *The Chemical Theatre of Biological Systems* (Eds. Hicks, M.G., Kettner, C.) Beilstein Symposium The Chemical Theatre of Biological Systems, May 2004, Beilstein-Institut, Logos-Verlag Berlin, pp 185 – 195.

[6]   Haynie, D.T. (2008) Biological Thermodynamics (2nd edition), Cambridge University Press.

[7]   Sievers, A., Beringer, M., Rodnina, M.V., Wolfenden, R. (2004) The ribosome as an entropy trap. *Proc. Natl. Acad. Sci. U.S.A.* **101**:7897 – 7901.
doi: 10.1073/pnas.0402488101.

[8]   Voet, D., Voet, J.G. (1995) Biochemistry (2nd edn.). Wiley, New York.

[9]   Atkinson, D.E. (1977) Cellular energy metabolism and its regulation. Academic Press, New York

[10]   Cornish-Bowden, A. (1983) Metabolic efficiency: is it a useful concept? *Biochem. Educ.* **11**:44 – 45.

[11]   Moffitt, J.R., Chemla, Y.R., Aathavan, K., Grimes, S., Jardine, P., Anderson, D.L.,
        Bustamante, C. (2009) Intersubunit coordination in a homomeric ring ATPase.
        Nature **457**:446 – 450.
        doi: 10.1038/nature07637

[12]   Soh, K.C., Hatzimanikatis, V. (2010) Network thermodynamics in the postgenomic
        Era. *Curr. Opinion Microbiol.* **13**:350 – 357.
        doi: 10.1016/j.mib.2010.03.001.

[13]   National Renewable Energy Laboratory: http://www.nrel.gov/.

[14]   Léger, C., Dementin, S., Bertrand, P., Rousset, M., Guigliarelli, B. (2004) Inhibition
        and aerobic inactivation kinetics of *Desulfovibrio fructosovorans* NiFe hydrogenase
        studied by protein film voltammetry. *J. Amer. Chem. Soc.* **126**:12162 – 12172.
        doi: 10.1021/ja046548d.

[15]   Saggu, M., Zebger, I., Ludwig, M., Lenz, O., Friedrich, B., Hildebrandt, P., Lendzian,
        F. (2009) Spectroscopic insights into the oxygen-tolerant membraneassociated [NiFe]
        hydrogenase of *Ralstonia eutropha* H16. *J. Biol. Chem.* **284**:1624 – 16276.
        doi: 10.1074/jbc.M805690200.

[16]   Wait, A.F., Parkin, A., Morley, G.M., dos Santos, L., Armstrong, F.A. (2010) Char-
        acteristics of enzyme-based hydrogen fuel cells using an oxygen-tolerant hydroge-
        nase as the anodic catalyst. *J. Phys. Chem.* C **114**:12003 – 12009.
        doi: 10.1021/jp102616m.

[17]   Pandelia, M.E., Fourmond, V., Tron-Infossi, P., Lojou, E., Bertrand, P., Léger, C.,
        Giudici-Orticoni, M.T., Lubitz, W. (2010) Membrane-bound hydrogenase I from the
        hyperthermophilic bacterium *Aquifex aeolicus*: enzyme activation, redox intermedi-
        ates and oxygen tolerance. *J. Amer. Chem. Soc.* **132**:6991 – 7004.
        doi: 10.1021/ja910838d.

[18]   Williams, R.J.P., Fraústo da Silva, J.J.R. (1996) The Natural Selection of the
        Chemical Elements. Clarendon Press, Oxford.

[19]   Le Goff, A., Artero, V., Joussel, B., Tran, P.D., Guillet, N., Métayé, R., Fihri, A.,
        Palacin, S., Fontecave, M. (2009) From hydrogenases to noble metal-free catalytic
        nanomaterials for H_2 production and uptake. *Science* **326**:1384 – 1387.
        doi: 10.1126/science.1179773.

[20]   Infossi, P., Lojou, E., Chauvin, J.P., Herbette, G., Brugna, M., Giudici-Orticoni, M.T.
        (2010) *Aquifex aeolicus* membrane hydrogenase for hydrogen biooxidation: role of
        lipids and physiological partners in enzyme stability and activity. *J. Hyd. Energ.*
        35:10778 – 10789.
        doi: 10.1016/j.ijhydene.2010.02.054.

[21] Lojou, E., Giudici-Orticoni, M.T., Bianco, P. (2005) Direct electrochemistry and enzymatic activity of bacterial polyhemic cytochrome c(3) incorporated in clay films. *J. Electroanal. Chem.* **579**:199 – 213.
doi: 10.1016/j.jelechem.2005.02.009.

[22] Lojou, E., Luo, X., Brugna, M., Candoni, N., Dementin, S., Giudici-Orticoni, M.T. (2008) Biocatalysts for fuel cells: efficient hydrogenase orientation for H_2 oxidation at electrodes modified with carbon nanotubes. *J. Biol. Inorg. Chem.* **13**:1157 – 1167.
doi: 10.1007/s00775-008-0401-8.

[23] Alonso-Lomillo, M.A., Rüdiger, O., Maroto-Valiente, A., Velez, M., Rodríguez-Ramos, I., Muñoz, F.J., Fernández, V.M., De Lacey, A.L. (2007) Hydrogenase-coated carbon nanotubes for efficient H_2 oxidation. *Nano Lett.* **7**:1603 – 1608.
doi: 10.1021/nl070519u.

[24] Luo, X.J., Brugna, M., Tron-Infossi, P., Giudici-Orticoni, M.T., Lojou, E. (2009) Immobilization of the hyperthermophilic hydrogenase from *Aquifex aeolicus* bacterium onto gold and carbon nanotube electrodes for efficient H_2 oxidation. *J. Biol. Inorg. Chem.* **14**:1275 – 1288.
doi: 10.1007/s00775-009-0572-y.

[25] Liebgott, P.P., Leroux, F., Burlat, B., Dementin, S., Baffert, C., Lautier, T., Fourmond, V., Ceccaldi, P., Cavazza, C., Meynial-Salles, I., Soucaille, P., Fontecilla-Camps, J.C., Guigliarelli, B., Bertrand, P., Rousset, M., Léger, C. (2010) Relating diffusion along the substrate tunnel and oxygen sensitivity in hydrogenase. *Nature Chem. Biol.* **6**:63 – 70.
doi: 10.1038/nchembio.276.

[26] Cornish-Bowden, A., Hofmeyr, J.-H.S., Cárdenas, M.L. (1995) Strategies for manipulating metabolic fluxes in biotechnology. *Bioorg. Chem.* **23**:439 – 449.
doi: 10.1006/bioo.1995.1030.

[27] Giudici, M.T. (2010) Metabolic studies of synthetic microbial ecosystems producing biohydrogen by dark fermentation. 9th International Hydrogenase Conference, Uppsala, Sweden.

Beilstein-Institut

Carbohydrate-Based Nanoscience: Metallo-glycodendrimers and Quantum Dots as Multivalent Probes

Raghavendra Kikkeri[1,2], Sung You Hong[1,2], Dan Grünstein[1,2], Paola Laurino[1,2] and Peter H. Seeberger[1,2,*]

[1]Department of Biomolecular Systems,
Max Planck Institute of Colloids and Interfaces,
Am Mühlenberg 1, 14424 Potsdam, Germany.

[2]Freie Universität Berlin, Institute of Chemistry and Biochemistry,
Arnimallee 22, 14195 Berlin, Germany.

E-Mail: *peter.seeberger@mpikg.mpg.de

Received: 3rd August 2010 / Published: 13th June 2011

Abstract

Rapid progress in nanoscience and its potential applications have spurred observers to predict that nanotechnology will the foremost science of the 21st century. Nanomaterials are beginning to have a major impact on research across the material and life sciences. While wide varieties of nanomaterials have been prepared with proteins, DNA, lipids and polymers, serious limitations arise with the neo-gly-coconjugates due to the ambiguous structures and lack of well-defined carbohydrates. This chapter highlights the contribution of glyconano-materials to biological, biochemical and biophysical studies. Particular focus will be placed on metallo-glycodendrimers and glyconanoparti-cles.

INTRODUCTION

Naturally occurring carbohydrates and oligosaccharides as well as glycoconjugates such as glycoproteins or glycolipids are present on the surface of nearly every cell within living systems [1]. These carbohydrates are known to have crucial roles in biological events as recognition sites between cells. They can trigger various phenomena such as cell growth, inflammatory responses or viral infections. In particular, the recognition phenomena between pathogens and host cells are thought to proceed *via* specific carbohydrate-protein interactions [2]. In comparison to the field of DNA/RNA (genomics) and proteins (proteomics), the understanding of the role of oligosaccharides/carbohydrates (glycomics) is very limited. While DNA-DNA and protein-protein interactions are well-studied and largely known, carbohydrate-carbohydrate and carbohydrate-protein interactions are poorly understood [1, 3]. The difficulties associated with studying carbohydrates are mainly a consequence of weak carbohydrate-protein interactions, the time-consuming synthesis and purification of oligosaccharides. In general, the interaction between a protein and a monosaccharide is weak with a dissociation constant (K_d) typically in the range of 10^{-4}—10^{-6} M compared with 10^{-6}—10^{-9} M antigen-antibodies interactions [4]. Multivalent presentation strongly enhances the binding to receptors on the cognate cells via polyvalent interactions, forming oligosaccharide structures of carbohydrate-protein receptors.

There is a clear need to develop new multivalent probes decorated with carbohydrates in order to investigate the information encoded by carbohydrates and biochemical events involving them. Such knowledge can open up novel platforms for a vast range of pharmaceutical, medical or biomedical commercial applications such as for imaging, targeted drug delivery, vaccine development and clinical diagnostics.

NANOPARTICLES AS MULTIVALENT PROBES

Nanoparticles are attractive multivalent systems: they exhibit unique pharmacokinetics (*e.g.* minimal renal filtration), very high surface to volume ratios, variation of material properties without change in chemical composition (*e.g.* quantum dot emission varies with size), they can exploit biological trafficking pathways (*e.g.* receptor-mediated endocytosis), and they have the potential for multi-functionality (*i.e.* both diagnostic and therapeutic) [5]. Multifunctional nanoparticles typically contain a particle 'core' and multiple surface moieties that can each endow the platform with distinct functionalities.

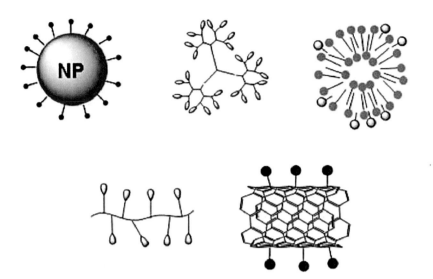

Figure 1. Selected multivalent carbohydrate probes: Top row from left to right, glyconanoparticle, glycodendrimer, glyco-functionalized liposome. Bottom row from left to right: glycopolymers and glycocarbon nanotubes.

Many multifunctional nanoparticle 'cores' have been explored including dendrimers, liposomes, gold, iron oxide, quantum dots, carbon nanotubes and degradable polymers (Fig. 1) [6]. Multivalent nanoparticles present several advantages over free carbohydrates. They offer the potential to improve targeting by combining lower affinity carbohydrates on the particle surface to increase the binding avidity over what can be achieved with a single carbohydrate. In addition, nanoparticles provide modularity whereby different glycan sequences or other compounds can be combined to customize therapy or show selectivity of targeting. Multivalency also provides additional sites for conjugation of polymers that improve nanoparticle pharmacokinetics (*e.g.* PEG). Finally, a multivalent surface offers multiple binding sites that increase the therapeutic 'payload' and carry combinations of therapeutics.

The initial forays of our group into this area employed metallo-glycodendrimers. Subsequently, we focused on the applications of quantum dots as a multivalent probe. Currently, many other systems are under evaluation.

MULTIVALENT PROBES: SUPRAMOLECULAR METALLO-GLYCODENDRIMERS

Multivalency is of fundamental importance for carbohydrate-protein interactions. Several methods have been developed to study those molecular interactions. However, multivalency, ligand placement, folding, and active adjustment of ligand positioning are topics that need to be addressed for the generation of polyvalent systems. Self-assembly processes are promising methods for the formation of such dendrimers, where electrostatic forces, hydrogen bonding, metal coordination and other non-covalent interactions control the assembly of the dendron [7]. Our laboratory initially explored the relevance of self-assembly processes to synthesize tunable fluorescent glycodendrimers. An amide derivative of 8-hydroxyquinoline confined glycodendrons was synthesized and coordinated with transition and lanthanide metal complexes upon self-assembly to obtain high nuclear glycodendrimers (Fig. 2).

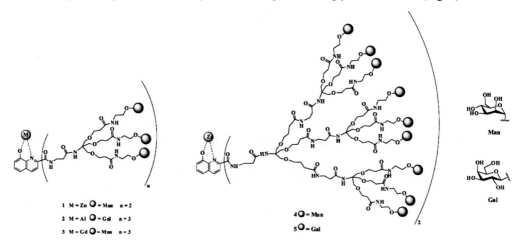

Figure 2. Glycodendrimers produced by self assembly.

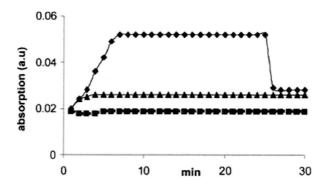

Figure 3. Turbidity analysis: absorption change of compound **1** (■), **5** (▲) and **4** (◆) at 500 nm on addition of ConA (1 mg ml⁻¹). Mannose (100 mM) was added to **4** after 25 min.

Complex formation was confirmed by various spectroscopic methods. Zn(II) and Al(III) complexes showed strong fluorescent intensities at 532 and 528 nm respectively and the quantum yields were approximately six to seven times higher than that of the dendrons. However, fluorescence of Gd(III) complexes were observed at 501 nm with almost the same quantum yield of the dendron due to weak HOMO-LUMO energy transfer of the Gd(III) metal ions. Finally, we have shown that high sugar density is essential for the lectin binding using a turbidity assay. The interaction of metallo-glycodendrimers with Concanavalin A (ConA) as a lectin showed high mannose sugar density depended turbidity (Fig. 3).

MULTIVALENT PROBES: RU(II)-GLYCODENDRIMERS

Inspired by the tunable and non-bleaching fluorescent nature of the metallo-glycodendrimers, we synthesized a series of Ru(II)-glycodendrimers and applied them to carbohydrate research. The Ru(II) core is most attractive for its octahedral core symmetry and robustness. Ru(II) complexes exhibit a low excited triple metal-to-ligand charge-transfer (^{3}MLCT) state and room temperature with ^{3}MLCT lifetimes of up to 1 µs. High emission quantum yields and strong oxidizing and reducing capabilities are important properties of theses complexes [8, 9].

Ru(II)-glycodendrimers are composed of three major portions: the inner shell, dendron branches and an outer shell (Fig. 4A). A Ru(II)(bipy)$_3$ metal complex is used as an inner core material to provide optical and electrochemical signal to the dendrimers and to control the geometry of the complex branched dendron that typically originates from a 4,4'-bipyridyl ligand of the inner core. Different branches result in different dendrimer generations. Finally, the outer shell is constituted with different carbohydrate moieties (Fig. 4B).

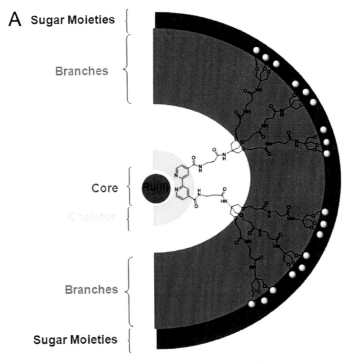

Figure 4. (A) Schematic representation of Ru(II)-glycodendrimer

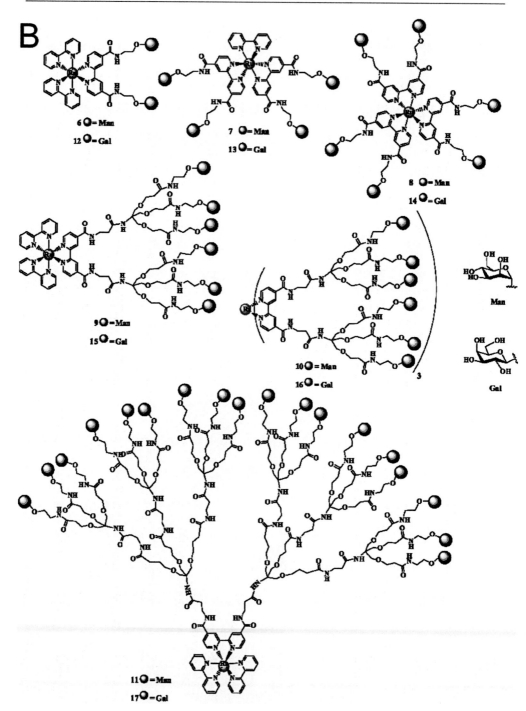

Figure 4. (B) Structures of Ru(II)-glycodendrimers **6-17**.

SYNTHESIS OF RU(II)-GLYCODENDRIMERS

The synthesis of Ru(II)-glycodendrimers relied on by using convergent method, where the core and dendrons were prepared separately and then united in the final step. Glycodendrons were prepared starting from tris-HCl **18**, following acylonitril addition, later treatment with conc. HCl in ethanol to yield tri-ester. Hydrolysis of triester, followed by coupling with pentafluorophenol afforded activated ester **19**. Pentafluorophenol ester was further reacted with peracetylated sugars that contain an anomeric 2-aminoethoxy linker to yield the final dendrons (**20** or **21**). Then, sugars were coupled with bipyridine derivatives and before the reaction with *cis*-Ru(bipy)$_2$Cl$_2$ resulted in complexes **9** and **11** respectively (Scheme 1).

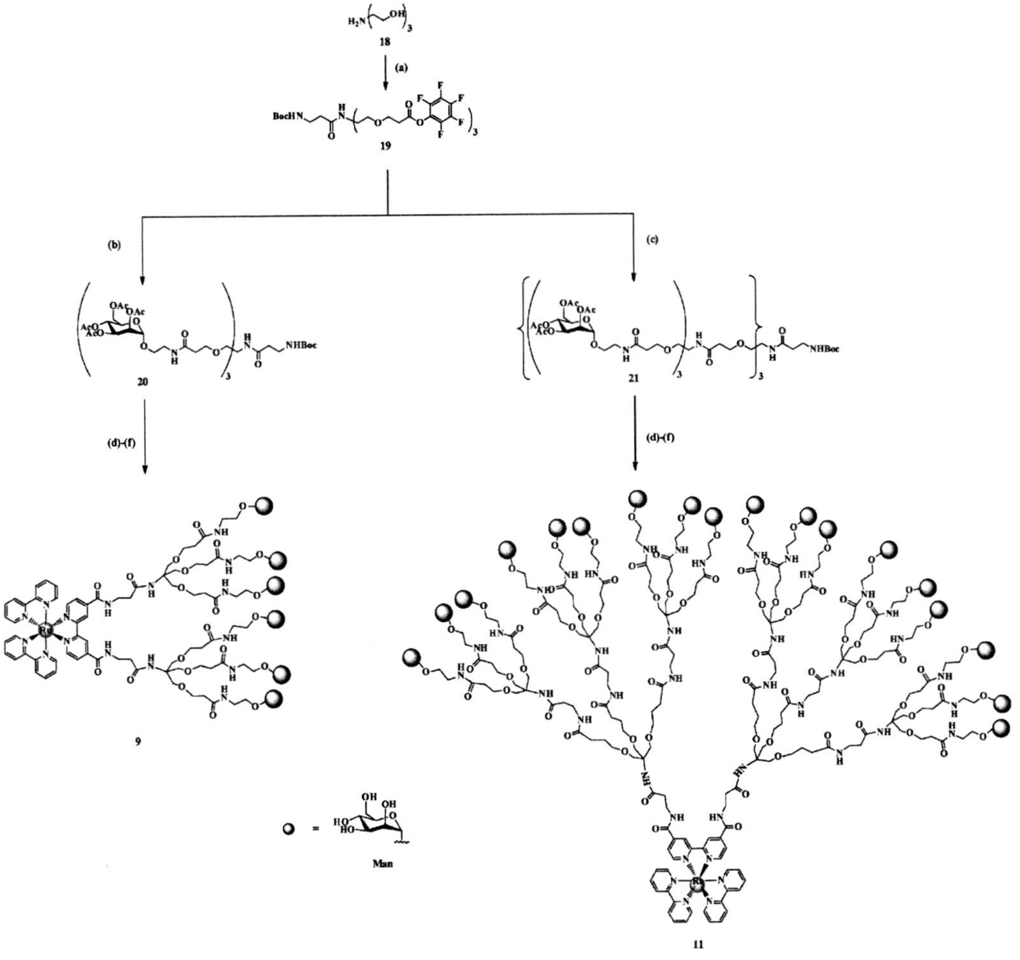

Scheme 1. Synthesis of Ru(II)-dendrimers **9** and **11**. (**a**) Acrylonitril/NaOH; Conc. HCl/ EtOH; *N*-Boc-β-Ala/DIC/HOBT/DCM; PFP-OH/DIC/HOBT/DCM; (**b**) 2-aminoethylα- D-peracetylated mannopyranoside/DCM; (**c**) TFA/**20**/DCM; (**d**) 2,2'-bipyridine 4,4'-di- carboxylic acyl chloride, DCM, TEA; (**e**) *cis*-Ru(bipy)$_2$Cl$_2$, EtOH; (**f**) NaOMe, MeOH.

We also devised a rapid and effective synthesis of Ru(II)-glycodendrimers bearing varying number of carbohydrates via Cu(I)-catalyzed [3+2] cycloaddition. Here, carbohydrate-dendrimers containing an azido-linker and Ru(II)-acetylene complexes were prepared separately. Subsequent Huisgen-[3+2] cycloaddition, followed by the removal of protecting groups on carbohydrate moieties provided access to the desired complexes in a straightforward and modular fashion (Scheme 2).

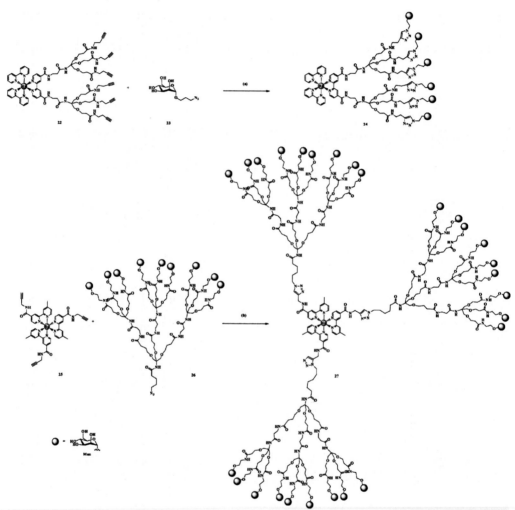

Scheme 2. Synthesis of Ru(II) complexes via Cu(I)-catalyzed [3+2] cycloaddition: (**a**) CuSO$_4$/ascorbic acid/THF:H$_2$O (1:1). (**b**) CuSO$_4$/ascorbic acid/THF:H$_2$O (1:1).

EFFECT OF CARBOHYDRATE DENSITY ON A RU(II)-GLYCODENDRIMER PROBE

After synthesizing Ru(II)-glycodendrimers with different carbohydrate moieties, we investigated how the carbohydrate density and bulk affected the photophysical properties and lectin binding affinity of these molecules in water. It is known that many fluorophores are quenched by water and the association with the dendritic structure might exhibit a shielding effect to reduce the quenching and increase of the quantum efficiency. As expected, the quantum yields of complexes **10** and **11** were approximately twice higher than those of the complexes **6-9**.

Table 1. Quantum yields of complexes **6-11**.

Entry	6	7	8	9	10	11
Quantum Yield	0.061	0.06	0.064	0.072	0.102	0.112

Ru(II) complexes are known to display long range energy and electron transfer processes. The rate of electron transfer in complexes **9-11** was investigated using photoinduced electron transfer (PET) between photo excited Ru(II)-templates and methyl viologen dication (MV^{2+}) as quencher. Complexes **9** to **11** showed almost an order of magnitude difference in quenching constant (K_q) and a decrease in life time. These results indicate that a high degree of carbohydrate density around the ruthenium core allows for efficient encapsulation and modification of the core properties.

Table 2. Photophysical data of Ru(II)-glycodendrimers. Quantum yield and life time were measured by excited complex **9-11** at $\lambda_{max} = 450$ nm and emission at $\lambda_{max} = 645$ nm.

Entry	λ_{max} [nm]	k_q [$M^{-1} \cdot s^{-1}$]	τ_o [μs]	I_o
9	645	$9.8 \cdot 10^8$	0.61	0.072
10	648	$1.8 \cdot 10^8$	1.31	0.102
11	648	$1.1 \cdot 10^8$	1.26	0.112
Ru(bipy)₃	613	$2.5 \cdot 10^9$	0.54	0.062

To evaluate the rate of energy transfer by **9-11**, we studied the formation of molecular oxygen in the singlet state upon photoexcitation of the ruthenium tris(bipyridine) complex. Tetramethyl piperidine (TEMP) was used as a trap for singlet oxygen to form a stable species (TEMPO) easily detected by EPR (see eq. 1) [9]. Continuous irradiation of Ru-complexes **9-11** in the presence of TEMP yielded a nitroxide triplet in the EPR spectrum. The rate of appearance of the TEMPO signal decreases from **9** to **11**, in support of the notion that carbohydrate encapsulation of the Ru(II)-template stops effective energy transfer to dissolved oxygen (Fig. 5).

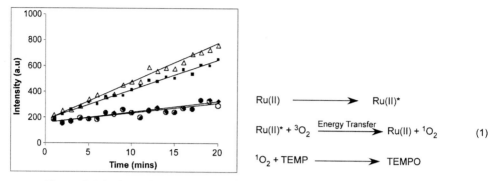

Figure 5. Kinetic profile of singlet oxygen formation upon irradiation of complexes **9** (■), **10** (◆), **11** (○) and Ru(bipy)₃ (△).

A Photoinduced Electron Transfer Based Lectin Sensor

After determining the photophysical properties of Ru(II)-glycodendrimers, we explored the donor/accetor concept of complexes **9-11**/BBV^{2+} interactions in lectin sensing process. Two lectins that recognize mannose were selected, ConA and galanthus nivilis agglutinin (GNA). Using a donor/acceptor mixture of complex **1** and BBV, a spontaneous gain in fluorescence was observed upon the addition of 75 nM of ConA and a further slow increase in the signal at 200 – 1000 nM was observed. In contrast, for 0 to 100 nM of ConA, complexes **10** and **11** displayed more modest gains in fluorescence compared to complex **9**, but a steady and linear increase upon the addition of 100 – 600 nM. Similar experiments with the higher valency lectin GNA were performed. The detection limits for the Ru-complexes were calculated based on these results (Table 3). Complex **9** is noticeably more sensitive than other sensors described in the literature[10].

Table 3. Detection limit of lectins using different Ru-mannose dendrimers

Compound	ConA [nM]	GNA [nM]
9	23 ± 3	25 ± 4
10	340 ± 12	328 ± 9
11	347 ± 14	331 ± 12

Optical Lectin Sensor

Ru(II) complexes are very also known as strong optical probes. To capitalize on this property, we developed a microarray based on direct carbohydrate-protein interaction read-out. ConA lectin was immobilized on a microarray prior to incubation with complexes **9**, **11**, **15** and **17**. Upon fluorescence scanning of the rinsed slides, strong fluorescent signals were observed on slides that were incubated with mannose complexes **9** and **11**. Using dendrimers **9** and **11** that contain six and eighteen mannoses respectively, ConA was detected at 0.125 mg/ml (620 nM).

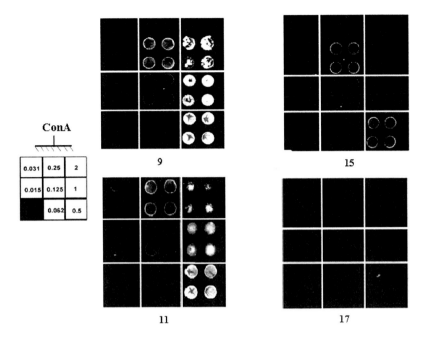

Figure 6. Incubation of Ru(II) dendrimers (**9**, **11**, **15** and **17**) with protein microarrays that contain different concentrations (mg/mL) of the lectin ConA (excitation at 480 nm).

ELECTROCHEMICAL LECTIN SENSOR

After establishing that Ru(II) glycodendrimers are useful tools to detected visually lectin-carbohydrate interactions, we utilized the redox the properties of Ru(II) core to develop electrochemical biosensor. ConA lectin was immobilized on a self-assembled monolayer on a gold surface prior to incubation of these surfaces with Ru(II)-complexes **9** and **11**. Following incubation, the chip was transferred to a electrochemical cell containing phosphate buffer. The scanning potential of 100 mV/s in the region of 1.0—1.4 V showed a peak at 1.62 μA. Repeated measurements at different time intervals revealed that maximal ConA/Ru(II)-complex interactions were reached after 240 min incubation (Fig. 7). Interestingly, incubation of complex **11** carrying 18 mannoses with ConA monolayers showed a very weak signal in the region of 1.0—1.4 V. An optimum current at 4.1 nA was obtained after 180 min incubation. Based on these findings, complex **9** was better suited for electrochemical sensing than the more complex dendrimer **11**. After establishing that the lectin-glycodendrimer interactions can be measured electrochemically, we determined the detection limit. Different concentrations of ConA were immobilized on gold substrates and treated with 0.5 mM of **9** prior to recording square-wave voltammetric (SWV) signals. At 2.5 nM the detection limit for **9** is comparable to other sensors [10].

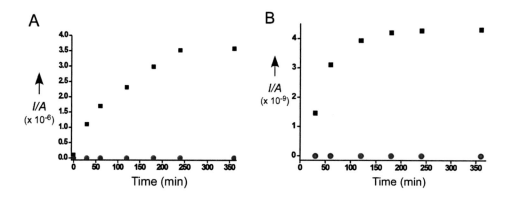

Figure 7. Square-wave voltammetric measurements at 1.14 V following incubation of **(A)** complexes **9** (■) and **15** (●) with ConA-functionalized surfaces for six hours; **(B)** complexes **11** (■) and **17** (●) with ConA-functionalized surfaces for six hours.

Table 4. Detection limits of ConA by different sensory systems

Methods	Detection Limits [nM]	References
Optical detection by Ru(II)-carbohydrate coated dendrimers and BBV by photoinduced electron transfer process	28 ± 3	[9a]
Optical detection by ConA microarray with Ru(II)-carbohydrate dendrimers	620	[9b]
Electrochemical detection by immobilizing ConA-Ru(II) dendrimers	2.5 ± 0.12	[9b]

REUSABLE SUGAR SENSOR

We have also developed a sugar sensor based on displacement of Ru(II)-glycodendrimer from the lectin-functionalized gold chips to allow for the detection of sugars that are bound by the lectin. ConA-functionalized gold chips containing **9** were immersed into solutions containing varying concentrations of D-glucose, D-mannose, α-D-man-(1→6)man, D-galactose, D-maltose or PIM glycans before SWV signals for Ru(II) were recorded. The current decreased in a concentration-dependent manner, indicating that the redox-active complex **9** is replaced in a competitive manner by the preferentially-binding carbohydrate. The detection limit for glucose (7 µM) compares favorably with the detection limits for other methods that are also in the micromolar range (Fig. 8) [11]. Similarly, other sugars resulted in different mode of current signal quenching and showed different detection limits (Table 5). Rapid quenching can be interpreted as a simultaneous displacement of weakly bound complex **9** from immobilized ConA and high affinity of the sugar for the lectin.

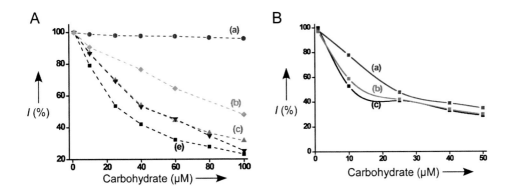

Figure 8. Response of square-wave voltammetric signals to increasing concentrations of **(A)** (a) D-galactose (●) (b) D-glucose (♦) (c) D-maltose (▲) (d) D-mannose (▲) and (e) D-manα(1 – 6)man (■); **(B)** (a) PIM3 (■); (b) Tri-mannose (■); (c) PIM4 (■).

Table 5. Detection limits of different free sugars by electrochemical ConA/Ru(II)-glycodendrimer method.

Methods	Detection limits (µM)
D-glucose	7 ± 0.12
D-mannose	3 ± 0.11
D-maltose	3 ± 0.06
D-galactose	–
D-manα(1 – 6)man	1.4 ± 0.12
PIM3	1.4 ± 0.11
Tri-mannose	0.61 ± 0.07
PIM4	0.61 ± 0.11

Primary level clinical diagnostic kits are expected to reset for repeated measurements. A gold chip exposed to 100 µM D-glucose solution was incubated with boronic acid substituted Merrifield resin to displace any sugar attached to the immobilized ConA. Incubation with complex **9** regenerated the surface for the next measurement. To verify the quality of the readings after regenerating the electrochemical detector, the chip was exposed to solutions containing 40 and 80 µM D-glucose. The platform was regenerated ten times using this re-iterative process (Fig. 9). The SWV signal decreases over the first six cycles and then remains constant for the last four regeneration cycles. Deactivation or effective hosting of glucose by ConA may be responsible for the observed decrease in the electrochemical signal after each cycle.

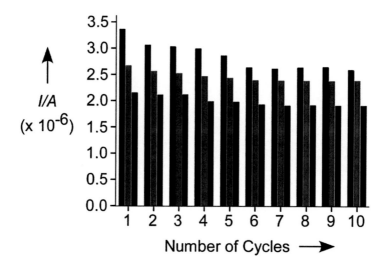

Figure 9. Maximum current signal upon regeneration of the ConA/**9** glucose detector: Complex **9** on gold substrate (black), addition of 40 µM of D-glucose (red), addition of 80 µM of D-glucose (blue).

MULTIVALENT PROBES: QUANTUM DOTS

Quantum dots are interesting multivalent tools to probe the glycome. There are different types of quantum dots, such as CdSe, CdS, or CdTe that are only a few nanometers in diameter and exhibit discrete size-dependent energy levels. As the size of the nanocrystal increases, the energy gap between HOMO-LUMO also increases, yielding a size-dependent rainbow of colours. Extensive tunability, from ultraviolet to infrared, can be achieved by varying the size and the composition of QDs, enabling simultaneous examination of multiple molecules and events. For example, small nanocrystals (~2 nm) made of CdSe emit in the range between 495 to 515 nm, whereas larger CdSe nanocrystals (~5 nm) emit between 605 and 630 nm [12].

QDs exhibit dramatically different properties when compared to organic fluorophores and fluorescent metal complexes. As illustrated in Table 6, organic and metal complex dyes typically have narrow absorption spectra and MLCT band, which means that they can only be excited within a narrow window of wavelengths. Furthermore, organic and metal complex dyes have broadened asymmetric emission spectra. In contrast, QDs have broad absorption spectra, enabling excitation by a wide range of wavelengths, and their emission spectra are symmetric and narrow. Moreover, QDs show superior quantum yield and multivalency compared to other fluorescent probes, and do not bleach [13].

Table 6. Photophysical properties of organic, metal fluorescent probe and quantum dots.

Optical properties	Organic fluorescent probe	Metal complexes	Quantum dots
Absorption band	Narrow	Narrow	Broad
Emission band	Broad	Broad	Narrow
Band tunabilty	Not good	Good	Good
Resistant to quenching	Not good	Good	Good
Photochemical stability	Not good	Good	Good
Emitting light intensity	Moderate	Good	High
Fixing ability to analysts	One-to-one	Multiple	Multiple

CARBOHYDRATES CONJUGATED TO QDS

The commercially available QDs are only soluble in nonpolar solvents because of their hydrophobic surface layer. For QDs to be useful probes for examination of biological specimens, the surface must be hydrophilic. Several strategies have been proposed to stabilize QDs in aqueous solutions. The easiest approach is to exchange the hydrophobic surfactant molecules with bifunctional molecules that are hydrophilic on one side and hydrophobic on the other side to bind to the ZnS shell. Most often, thiols (-SH) are used as anchoring groups on the ZnS surface and amine groups are used as the hydrophilic ends. Recently, dithols such as DL-thioctic acid have been used to prepare PEG linkers **29** and **33**. PEG groups were used to avoid non-specific interaction by the ZnS surface. We used different length PEG linkers and found that PEG_{2000} is the best model to study carbohydrate interactions. The PEGylated QDs **30** were further treated with 2-*N*-hydroxy succinimide maleimido linker to obtain QD-PEG-maleimido **31**. Finally, the QD-PEG-maleimido was reacted with thio-sugars to obtain final compounds **32**.

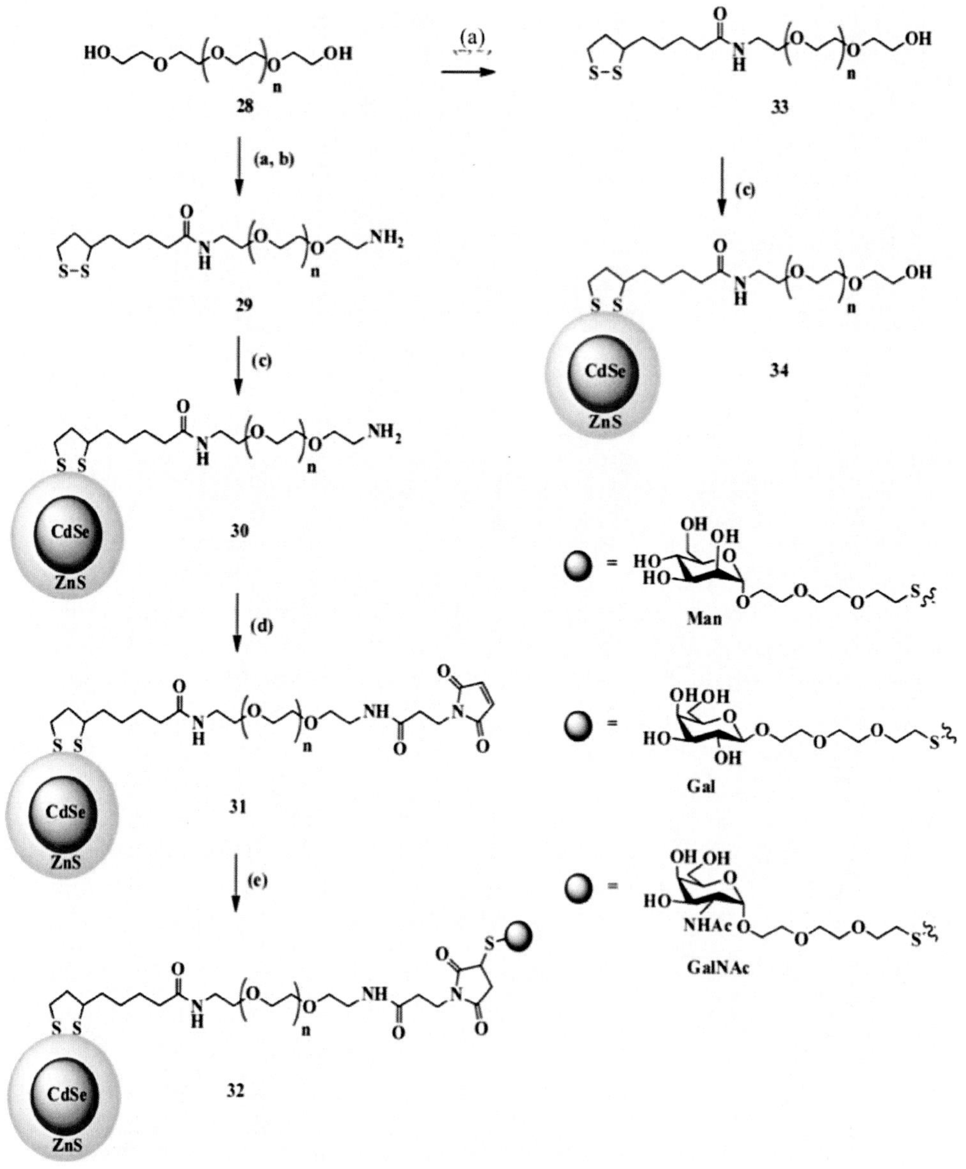

Kikkeri, R. *et al.*

Scheme 3. Synthesis of QDs: **(a)** MsCl, TEA, NaN₃, 12 h; Ph₃P, H₂O, 12 h; **(b)** Thioctic acid, DIC, NHS, 12 h; **(c)** NaBH₄, MeOH/H₂O; CdSe/ZnS, EtOH; **(d)** 4-maleimidopropanoic acid NHS ester, pH 8.5; **(e)** Man-SH or Gal-SH or GalNH₂-SH, pH 7.5.

IN VIVO AND IN VITRO EXPERIMENTS WITH QDS

Fluorescent probes are widely used in cell biology for probing structure and to locate specific receptors. Owing to their robust optical properties, QDs are ideal probes in this area. Here, we employed HepG2 cells that express asialoglycoprotein receptor (ASGP-R) that bind to galactose glycoproteins. HepG2 cells were cultured in presence of QDs coated with galactose sugars. Flow cytometry (Fig. 10) after 2 h of incubation of the cells with QDs revealed that Gal- capped QDs were taken up by the HepG2 cells preferentially over PEG$_{2000}$-capped QDs **34.**

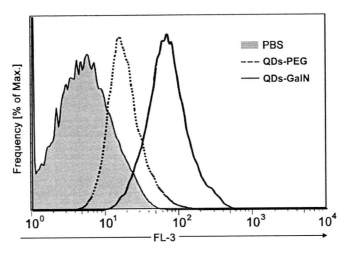

Figure 10. Specific uptake of D-GalN-capped QDs by HepG2 cells. HepG2 cells were incubated overnight with 20 nmol of PEG$_{2000}$ QDs (dashed line) or 20 nmol of GalN-PEG$_{2000}$ QDs (solid line). As negative control, PBS was added to the cells.

After demonstrating specific uptake of D-GalN-capped QDs *in vitro* we analyzed specific targeting of the liver *in vivo*. For this purpose, mice received either PEG$_{2000}$-QDs or QDs capped with D-mannose or D-galactosamine by intravenous (*i.v.*) injection (PBS buffer was injected as a negative control). A low level of unspecific sequestration was observed in the liver 2 h after injection of PEG$_{2000}$-capped QDs (Fig. 11). In contrast, injection of Man-PEG$_{2000}$ and also GalN-PEG$_{2000}$ capped QDs resulted in a high number of QDs sequestering in the liver. This finding suggests binding and/or endocytosis of the QDs mediated by mannose receptor and ASGP-R. ASGP-R is expressed predominantly on hepatocytes, while the mannose receptor is strongly expressed on Kupffer cells and sinusoidal endothelial cells in the liver. This finding indicates that carbohydrate-protein interactions exhibit specificity and may be exploited for targeted drug delivery *in vivo*.

160

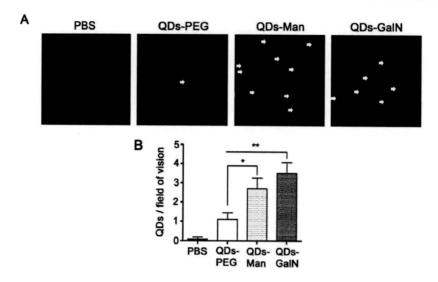

Figure 11. Specific liver sequestration of D-Man and D-GalN capped QDs in liver. **(A)** Paraffin sections of the livers were prepared, and QD sequestration in the liver was visualized by fluorescence microscopy. Arrows indicate QDs sequestered to liver tissue. **(B)** Statistical analysis of QD sequestration in the liver was performed by counting 10 microscopic fields of vision for each mouse. Data are presented as mean (SEM for each group (*$P < 0.05$, **$P < 0.01$).

CONCLUSION

In conclusion, our laboratory has developed a metallo-glycodendrimer and quantum dots based novel platform to study carbohydrate-protein interactions. The optical and electrochemical signals from Ru(bipy)$_3$ and CdSe core offers direct readout techniques to image and analyze specific interactions. Additionally, Cu(I)-catalyzed [3+2] cycloaddition and PEGylated sugar conjugation reactions present the opportunity to minimize sugar consumption. The development of sensitive lectin and sugar sensors in optical and electrochemical mode will enable glycobiologists to screen large numbers of carbohydrates that are thought to have important roles in biological systems. Even though impressive applications of quantum dots and metallo-glycodendrimers have been reported, some major drawbacks associated with the biocompatibility and stability of nanoparticles still need to be addressed. Thus, the choice of appropriate nanomaterials for medical applications is still a land of opportunity for material scientists, chemists and glycobiologists.

ACKNOWLEDGEMENTS

We gratefully acknowledge generous funding from the Max Planck Society.

REFERENCES

[1]   (a) Werz, D.B., Seeberger, P.H. (2000) *Nat. Rev.* **4**:751.
      (b) Angata, T., Varki, A. (2000) *Chem. Rev.* **102**:439.
      doi: 10.1021/cr000407m
      (c) Varki, A., Cummins, R., Esko, J., Freeze, H., Hart, G., Marth, J. (Eds.) (1999)
      Essentials of Glycobiology. Consortium of Glycobiology Editors, La Jolla, Califor-
      nia. Cold Spring Harbor Laboratory Press, Cold Spring Harbor, NY.

[2]   (a) Ada, G., Issacs, D. (2003) *Clin. Microbiol. Infect.* **9**:79.
      doi: 10.1046/j.1469-0691.2003.00530.x
      (b) Verez-Bencomo, V. (2004) *Science* **305**:522.
      doi: 10.1126/science.1095209
      (c) Hecht, M.-L., Stallforth, P., Varon Silva, D., Adibekian, A., Seeberger, P.H.
      (2009) *Curr. Opin. Chem. Biol.* **13**:354.
      doi: 10.1016/j.cbpa.2009.05.127
      (d) Vliegenthart, J.F.G. (2006) *FEBS Lett.* **580**:2945.
      doi: 10.1016/j.febslet.2006.03.053
      (e) Landers, J.J., Cao, J., Lee, I., Piehler, L.T., Myc, P.P., Myc, A., Hamouda, T.,
      Galecki, A.T., Baker, J.R. (2002) *Infect. Dis.* **186**:1222.
      doi: 10.1086/344316
      (f) Heidecke, C.D., Lindhorst, T.K. (2007) *Chem.-Eur. J.* **13**:9056.
      doi: 10.1002/chem.200700787

[3]   (a) Caruthers, M.H. (1985) *Science* **230**:281.
      doi: 10.1126/science.3863253
      (b) Merrifield, R.B. (1985) *Angew. Chem. Int. Ed.* **24**:799.
      doi: 10.1002/anie.198507993

[4]   (a) Liang, R., Yan, L., Loebach, J., Ge, M., Uozumi, Y., Sekanina, L., Horan, N.,
      Gildersleeve, J., Thompson, C., Smith, A. (1996) *Science* **274**:1520.
      doi: 10.1126/science.274.5292.1520
      (b) Liang, R., Loebach, J., Horan, N., Ge, M., Thompson, C., Yan, L., Kahne, D.
      (1997) *Proc. Natl. Acad. Sci. U.S.A.,* **94**: 10554
      doi: 10.1073/pnas.94.20.10554

[5]   Ruoslahti, E. (2002) *Cancer Cell,* **2**:97.
      doi: 10.1016/S1535-6108(02)00100-9.

[6]   (a) Harisinghani, M.G., Weissleder, R. (2004) *PLoS Med.* **1**:202.
      doi: 10.1371/journal.pmed.0010066.
      (b) Choi, Y., Baker, J.R. (2005) *Cell Cycle* **4**:669.
      doi: 10.4161/cc.4.5.1684.

(c) Patri, A.K., Myc, A., Beals, J., Thomas, T.P., Bander, N.H., Baker, J.R. (2004) *Bioconj. Chem.* **15**:1174.doi: 10.1021/bc0499127.

(d) Quintana, A., Raczka, E., Piehler, L., Lee, I., Myc, A., Majoros, I., Patri, A.K., Thomas, T., Mule, J., Baker, J.R. (2002) *Pharmaceut. Res.* **19**:1310. doi: 10.1023/A:1020398624602.

(e) Torchilin, V.P., Levchenko, T.S., Lukyanov, A.N., Khaw, B.A., Klibanov, A.L., Rammohan, R., Samokhin, G.P., Whiteman, K.R. (2001) *Biochim. Biophys. Acta - Biomembranes* **1511**:397. doi: 10.1016/S0005-2728(01)00165-7 .

(f) Kakudo, T., Chaki, S., Futaki, S., Nakase, I., Akaji, K., Kawakami, T., Maruyama, K., Kamiya, H., Harashima, H. (2004) *Biochemistry* **43**:5618. doi: 10.1021/bi035802w.

(g) Mastrobattista, E., Crommelin, D. J. A., Wilschut, J., Storm, G. (2002) *J. Liposome Res.* **12**:57. doi: 10.1081/LPR-120004777.

(h) O'Neal, D.P., Hirsch, L.R., Halas, N.J., Payne, J.D., West, J.L. (2004) *Cancer Lett.* **209**:171.

(i) Hirsch, L.R., Stafford, R.J., Bankson, J.A., Sershen, S.R., Rivera, B., Price, R.E., Hazle, J.D., Halas, N.J., West, J.L. (2003) *Proc. Natl. Acad. Sci. U.S.A.* **100**:13549. doi: 10.1073/pnas.2232479100.

(j) Hainfeld, J.F., Slatkin, D.N., Smilowitz, H.M. (2004) *Phys. Med. Biol.* **49**:N309. doi: 10.1088/0031-9155/49/18/N03.

[7]   (a) Gomez-Garcia, M., Benito, J.M., Rodriguez-Lucena, D., Yu, J.-X., Chmurski, K., Mellet, C.O., Gallego, R.G., Maestre, A., Defaye, J., Fernandez, J.M.G. (2005) *J. Am. Chem. Soc.* **127**:7970. doi: 10.1021/ja050934t.

(b) Mellet, C.O., Defaye, J., Fernandez, J.M.G. (2002) *Chem.-Eur. J.* **8**:1982. doi: 10.1002/1521-3765(20020503)8:9<1982::AID-CHEM1982>3.0.CO;2-5.

(c) Fulton, D.A., Stoddart, J.F. (2001) *Bioconj. Chem.* **12**:655. doi: 10.1021/bc0100410.

(d) Lin, C.-C., Yeh, Y.-C., Yang, C.-Y., Chen, C.-L., Chen, G.-F., Chen, C.-C., Wu, Y.-C. (2002) *J. Am. Chem. Soc.* **124**:3508. doi: 10.1021/ja0200903.

(e) Otsuka, H., Akiyama, Y., Nagasaki, Y., Kataoka, K. (2001) *J. Am. Chem. Soc.* **123**:8226. doi: 10.1021/ja010437m.

(f) Kawa, M., Frechet, J.M.J. (1998) *Chem. Mater.* **10**:286. doi: 10.1021/cm970441q.

(g) Blasini, D.R., Flores-Torres, S., Smilgies, D.-M., Abruna, H.D. (2006) *Langmuir* **22**:2082. doi: 10.1021/la052558w.

(h) Elizarov, A.M., Chang, T., Chiu, S.-H., Stoddard, J.F. (2002) *Org. Lett.* **4**:3565.
doi: 10.1021/ol026479c.

(i) Gibson, H.W., Yamaguchi, N., Hamilton, L., Jones, J.W. (2002) *J. Am. Chem. Soc.* **124**:4653.
doi: 10.1021/ja012155s.

(j) Kamiya,N., Tominaga, M., Sato, S., Fujita, M. (2007) *J. Am. Chem. Soc.* **129**:3816.
doi: 10.1021/ja0693082.

[8]   (a) Pollak, K.W., Leon, J.W., Frechet, J.M.J., Maskus, M., Abruna, H.D. (1998) *Chem. Mater.* **10**:30.
doi: 10.1021/cm970312+.

(b) Weyermann, P., Gisselbrecht, J.-P., Boudon, C., Diederich, F., Gross, M. (1999) *Angew. Chem. Int. Ed.* **38**:3215.
doi: 10.1002/(SICI)1521-3773(19991102)38:21<3215::AID-ANIE3215>3.0.CO;2-S.

(c) Larsen, J., Bruggemann, B., Khoury, T., Sly, J., Crossley, M.J., Sundstrom, V., Akesson, E. (2007) *J. Phys. Chem. A* **111**:10589.
doi: 10.1021/jp070545g.

[9]   (a) Kikkeri, R., Garcia-Rubio, I., Seeberger, P. H. (2009) *Chem. Commun.* **235**.
doi: 10.1039/b814146k.

(b) Kikkeri, R., Kamena, F., Gupta, T., Hossain, L.H., Boonyarattanakalin, S., Gorodyska, G., Beurer, E., Coullerez, G., Textor, M., Seeberger, P.H. (2010) *Langmuir*, **26**:1520.
doi: 10.1021/la9038792.

(c) Ibey, B.L., Beier, H.T., Rounds, R.M., Cote, G.L., Yadavalli, V.K., Pishko, M.V. (2005) *Anal. Chem.* **77**:7039.
doi: 10.1021/ac0507901.

(d) Krist, P., Vannucci, L., Kuzma, M., Man, P., Sadalapure, K., Patel, A., Bezouska, K., Pospisil, M., Petrus, L., Lindhorst, T.K., Kren, V. (2004) *ChemBioChem* **5**:445.
doi: 10.1002/cbic.200300669.

(e) Okada, T., Makino, T., Minoura, N. (2009) *Bioconj. Chem.* **20**:1296.
doi: 10.1021/bc900101u.

(f) Hasegawa, T., Yonemura, T., Matsuura, K., Kobayashi, K. (2003) *Bioconj. Chem.* **14**:728.
doi: 10.1021/bc020026a.

(g) Hasegawa, T., Yonemura, T., Matsuura, K., Kobayashi, K. (2001) *Tetrahedron Lett.* **42**:3989.
doi: 10.1016/S0040-4039(01)00424-5.

(h) Kikkeri, R., Liu, X.Y., Adibekian, A., Tsai, Y.H., Seeberger, P.H. (2010) *Chem. Commun.* **46**:2197.
doi: 10.1039/b925113h.

(i) Gottschaldt, M., Schubert, U.S., Rau, S., Yeno, S., Vos, J.G., Kroll, T., Clement, J., Hilger, I. (2010) *ChemBioChem* **11**:649.
doi: 10.1002/cbic.200900769.

[10]   (a) Guo, C., Boullanger, P., Jiang, L., Liu, T. (2007) *Biosens. Bioelectron.* **22**:1830.
doi: 10.1016/j.bios.2006.09.006 .
(b) Huang, C.-C., Chen, C.-T., Shiang, Y.-C., Lin, Z.-H., Chang, H.-T. (2009) *Anal. Chem.* **81**, 875.
doi: 10.1021/ac8010654.

[11]   (a) Takahashi, S., Anzai, J. (2005) *Sens. Lett.* **3**:244.
doi: 10.1166/sl.2005.026.
(b) Sato, K., Kodama, D., Anzai, J.-I. (2006) *Anal. Bioanal. Chem.* **386**:1899.
doi: 10.1007/s00216-006-0779-5.
(c) Sato, K., Imoto, Y., Sugama, J., Seki, S., Inoue, H., Odagiri, T., Hoshi, T., Anazai, J.-I. (2005) *Langmuir* **21**:797.
doi: 10.1021/la048059x.
(d) Xiao, Y., Patolsky, F., Katz, E., Hainfeld, J.F., Willner, I. (2003) *Science* **299**:1877.
doi: 10.1126/science.1080664.
(e) Zayats, M., Katz, E., Baron, R. (2005) *J. Am. Chem. Soc.* **127**:12400.
doi: 10.1021/ja052841h.
(f) Bahshi, L., Frasconi, M., Telvered, R., Yehezkeli, O., Willner, I. (2008) *Anal. Chem.* **80**:8253.
doi: 10.1021/ac801398m.
(g) Malitesta, C., Losito, L., Zambonin, P.G. (1999) *Anal. Chem.* **71**:1366.
doi: 10.1021/ac980674g.
(h) Jazkumar, D.R.S., Narayanan, S.S. (2009) *Carbon* **47**:957.
doi: 10.1016/j.carbon.2008.11.050.

[12]   (a) Giepmans, B.N., Adams, S.R., Ellisman, M.H., Tsien, R.Y. (2006) *Science* **312**:217.
doi: 10.1126/science.1124618.
(b) Liu, W., Choi, H.S., Zimmer, J.P., Tanaka, E., Frangioni, J.V., Bawendi, M. (2007) *J. Am. Chem. Soc.* **129**:14530.
doi: 10.1021/ja073790m.
(c) Clapp, A.R., Medintz, I.L., Uyeda, H.T., Fisher, B.R., Goldman, E.R., Bawendi, M.G., Mattoussi, H. (2005) *J. Am. Chem. Soc.* **127**:18212.
doi: 10.1021/ja054630i.
(d) Clapp, A.R., Medintz, I.L., Mauro, J.M., Fisher, B.R., Bawendi, M.G., Mattoussi, H. (2004) *J. Am. Chem. Soc.* **126**:301.
doi: 10.1021/ja037088b.

(e) Chan, W.C.W., Nie, S. (1998) *Science* **281**:2016.
doi: 10.1126/science.281.5385.2016.

[13]   Guo, W., Li, J.J., Wang, Y.A., Peng, X. (2003) *Chem. Mater.* **15**:3125.
doi: 10.1021/cm034341y.

DESIGN OF HIERARCHICALLY SCULPTURED BIOLOGICAL SURFACES WITH ANTI-ADHESIVE PROPERTIES

KERSTIN KOCH

Rhine-Waal University of Applied Science, Faculty of Life Sciences,
Landwehr 4, 47533 Kleve, Germany

E-MAIL: koch@hochschule-rhein-waal.de

Received: 14th September 2010 / Published: 13th June 2011

ABSTRACT

In many plant species sophisticated functions such as water repellence, reduction of particle adhesion and reduction of insect attachment are correlated to a hierarchically sculptured surface design. One prominent example is given by the hierarchically sculptured, self-cleaning surface of the lotus leave (*Nelumbo nucifera*). In plants hierarchy of surfaces is often realized by combining microstructures with superimposed self-assembled nanostructures. Such functional biological surfaces are of great interest for the development of biomimetic self-cleaning materials. Examples of superhydrophobic plant surfaces are introduced here, and their hierarchical surface sculptures and existing and potential use of their properties in artificial materials are shown.

INTRODUCTION

The term *biomimetic* describes the study and transfer of nature's methods, mechanisms and processes into artificial materials and systems [1]. Biomimetic research deals with functional micro- and nanostructures and the transmission of biological principles such as self-organization and functional structures. To use nature's models in engineering one can take biological characteristics and develop analogue solvents for engineering. Several aspects, such as their energy harvesting system, their ability of self-healing and their multifunctional surface properties make plants interesting models in biomimetic research.

Approximately 460 million years ago, the first plants moved from their aqueous environment to the drier atmosphere on land. Since then, evolutionary processes led to physiological, chemical and morphological variations which enabled them to settle into nearly all conceivable habitats. The plant surface is in direct contact with the environment and therefore an important interface with different environmental influences. Adapted to their specific environmental approaches, *e. g.* efficient light reflection in desert plants, a large diversity of functional plant surface structures has evolved [2].

The epidermis is the outermost cell layer of the primary tissues of all leaves and several other organs of plants. The protective outer coverage of the epidermis is a continuous extracellular membrane, called the cuticle. The cuticle is basically composed by a polymer called cutin and integrated and superimposed lipids called "waxes" [3]. Only a few books summarize the intensive research and include further aspects, such as the biosynthesis of the plant cuticle [4, 5]. One of the most important properties of the cuticle is the transpiration barrier function, *e. g.* for the reduction of water loss and prevention of leaching of ions from the inside of the cells to the environment. The plant cuticle also plays an important role for insect and microorganism interaction. Their surface sculptures have a strong influence on the reflection of radiation, the surface wetting ability and adhesion of insects and contaminations [2]. Additionally, it is a mechanical stabilization element for the plant tissue [6].

Waxes are the hydrophobic component of the plant cuticle and are integrated (intracuticular) and superimposed (epicuticular) to the cuticle [3]. The epicuticular waxes form thin two-dimensional films and three-dimensional structures which cover the cuticle surface. Wax films are often combined with three-dimensional waxes, which occur in different morphologies like tubules, rodlets or platelets (Fig. 1) [7]. Both the films and the three-dimensional waxes have a crystalline structure [8]. Three-dimensional epicuticular wax structures occur in sizes from 0.5 to 100 µm, whereas two-dimensional wax films range from a few molecular layers up to 0.5 µm [7, 9]. The wax morphology originates by self-assembly [10] and is strongly correlated to the wax chemical composition [3, 11]. The chemistry of waxes is a mixture of long-chain hydrocarbons and, in some waxes, cyclic hydrocarbons. Substitution by functional groups (-hydroxyl, -carboxyl, -ketoyl) broadens the spectrum of compounds to fatty acids, aldehydes, β-diketones and primary and secondary alcohols [3, 12]. The chemical composition of plant waxes is highly variable amongst plant species, the organs of one species, and varies during organ ontogeny [13].

Figure 1. Scanning electron microscopy graph of epicuticular waxes on a cabbage leaf (*Brassica oleracea*) shows perpendicular orientated rodleds and interspersed some smaller wax filaments.

HIERARCHY IN PLANT SURFACES

Hierarchy of the plant surface sculptures is based on the combination of sculptural elements of different scale sizes. In the following examples the defined level of hierarchy start at the macro-scale and are subdivided into further hierarchical levels with smaller sculptural elements. The sculptural subunits in the macro-dimension range in the sizes of some milli-metres, *e. g.* the waviness of a leaf plane. Multi-cellular sculptures, such as the hairs shown in Figure 3, can create a second level of sculptural elements. Within the next smaller level, which covers sculptures of several microns, the outline of surface cells creates a sculpturing. The outline of single cells can be convex (arced to the outside) or concave (arced to the inside). The convex cell type is the most common one and is often found on flower-leaves, stems and leaves [14]. A description of different cell morphologies and cell outlines is given by Barthlott and Ehler [15] and Koch *et al.* [2]. Within the next smaller level, which covers sculptures of up to a few micrometer down to sub-micrometer scale sizes, sculptures, such as cuticular folds and the wax crystals are relevant.

HIERARCHICAL SURFACE SCULPTURES FOR SUPERHYDROPHOBICITY AND SELF-CLEANING

Hierarchical sculptures play a key role in surface wetting [16] and are discussed in this paper in the context of superhydrophobicity. Superhydrophobic surfaces play an important role in technical applications ranging from self-cleaning window glasses, paints, and textiles [17] and include low-friction surfaces for fluid flow [18] and energy conservation [19]. Such superhydrophobic surfaces are characterized by a high static contact angle (> 150°), and in the case of self-cleaning surfaces also show a low contact angle hysteresis (the difference between the advancing and receding contact angles) of less than 10° and a low tilting angle (< 10°).

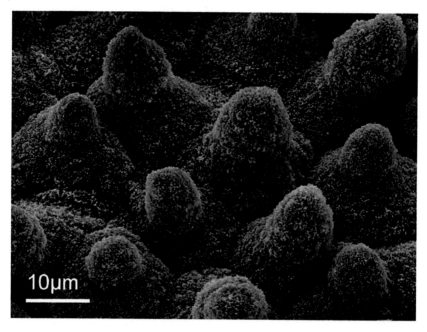

Figure 2. Scanning electron micrograph of the lotus leaf surfaces shows the papilla shaped cells, covered a nanostructure of wax tubules.

The most prominent self-cleaning surfaces in nature are the lotus leaves (*Nelumbo nucifera*). On lotus leaves, water droplets roll over the leaf surface and collect dirt and other particles from the surfaces. Lotus leaves have been the inspiration for the development of several artificial self-cleaning biomimetic materials [17, 20]. The superhydrophobicity and self-cleaning of the lotus leaves (Fig. 2) were found to be a result of a hierarchical surface structure, built by randomly oriented small hydrophobic wax tubules on the top of convex cell papillae [21, 22]. The wax tubules on lotus leaves, shown in Figure 2, are usually 0.3 to 1.1 µm in length and 0.1 to 0.2 µm in diameter [11]. In most plant species, superhydrophobic surfaces are designed by a two-level surface sculpturing: convex (papillose) sculptured cells covered with three-dimensional waxes [23]. In different studies, artificial superhydrophobic surfaces with surface sculpturing only on the nano-scale level have been fabricated [24], but hierarchy in surface sculpturing has been shown to further reduce the adhesion (hysteresis) of liquids and water droplets roll off at very low tilt angles, a condition required for self-cleaning [20, 21, 23, 25, 26–28]. A plausible explanation for these phenomena can be found in Wentzel's explanations [29]. Wentzel suggested a simple model predicting that the wetting ability of a liquid, measured as the contact angle, with a rough surface is different from that with a smooth surface and that a hydrophobic surface gets more hydrophobic with the increase of the surface roughness. Later, Cassie and Baxter [30] showed that a gaseous phase, commonly referred to as "air", may be trapped in the cavities of a rough surface, resulting in a composite solid-liquid-air interface. Nowadays, a large number of studies give evidence that water repellence and self-cleaning of superhydrophobic surfaces are caused by a reduction of both the contact area and the adhesion of contaminations and water to the surface [31, 32].

As mentioned above, the lotus leaf surface has been termed a two-level hierarchical surface. Recently, cryo-scanning electron microscopy showed that the minimized contact area of a lotus leaf and an applied liquid (a glycerol-water mixture) is caused by a four-level surface sculpturing [33]. At the scale of several hundreds of microns the waviness of the lotus leaf surface causes large air pockets, in which several cells are only in contact with the air captured under the water droplet. The second hierarchical level is given by the variations in epidermal cell heights. In those areas where the leaf surface is in contact with the liquid, only the higher cells are in contact with the liquid. The third level of hierarchy, which has an influence on the contact area, is provided by the clusters formed by the wax tubules. Such clusters are randomly interspersed within the surface and they are higher than the surrounding wax tubules. The fourth level of hierarchy which reduces the water-surface contact area is given by the single tubules emerging out of the wax clusters [34].

Hierarchy in Superhydrophobic Surfaces for Underwater Use

Underwater air-retaining surfaces are of great technological, economic and ecological interest because the trapped air minimizes the water-solid contact area and leads to a reduction of frictional drag between surface and water [35, 36]. Floating plants of the genus *Salvinia* possess superhydrophobic leaf surfaces which are adapted to efficiently retain a layer of air when submerged under water [18]. Immersed in water the leaves are capable of holding an air layer for several weeks. The upper sides of the floating leaves of *Salvinia* are covered with multicellular hairs (Fig. 3). Water applied onto the leaf stays on top of the hairs without sinking in between the structures and the water is not able to penetrate between the hairs; thus forming an air-water interface between the tips of the hairs and the applied water. For long term air retention the stabilization of this air-water interface at a predefined level at the top of the hairs is crucial. Scanning electron microscopy studies of *Salvinia molesta* leaves showed that six levels of surface sculptures exist. The first level is given by the arrangement of several multicellular hairs (also known as "eggbeater" hairs) within the leaf plane. The individual hairs, variations in their sizes and their orientation provide the second level of surfaces sculpturing. The convex epidermis cells of the hairs are the third level of sculpturing.

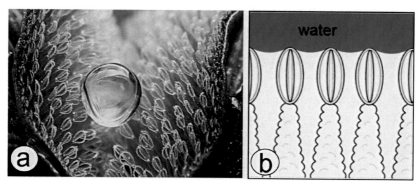

Figure 3. The water fern Salvinia as model for underwater air retaining surfaces. **(a)** shows a water droplet on the superhydrophobic leaf of the water fern *Salvinia molesta*. In **(b)** the schematic shows the eggbeater shaped hairs of *Salvinia* surrounded by air, when immersed under water.

The fourth level is given by the three-dimensional wax crystals, which appear in two different sizes and morphologies. The larger wax rodlets provide the fourth level, whereas the fifth level of sculpture is given by smaller wax filaments. The wax rodlets have a height of approximately 300 nm and the height of the small wax filaments between the rodlets is approximately 30 nm. Finally, the wax rodlets possess another level of surface sculpturing formed by small filament-like structures. The terminal cells of each hair lack the wax crystals and are in contrast to the rest of the upper leaf surface not superhydrophobic but hydrophilic. These hydrophilic patches pin the air-water interface to the tips of the hairs and thus decrease the risk of delamination of the water layer. For artificial air-retaining surfaces, Barthlott *et al.* [18] expect a wide range of applications, including drag-reducing ship coatings, low-friction fluid transport and novel concepts for thermally insulating interfaces.

SLIDING STRUCTURES FOR INSECT CAPTURE

The examples introduced here show that multilevel hierarchically surface sculpturing can provide water repellent, self-cleaning or underwater air-retaining surfaces with drag-reducing properties. Hierarchy in surface sculpturing can also play a key role in anti-adhesive surfaces. Poppinga *et al.* [37] investigated by scanning electron microscopy the surface sculptures of 53 different plants species with pitfall traps for capturing insects. Two groups of flowering plants specialized in insect catching were investigated. The carnivorous plants catch prey for a substantial nutrient supply [38] and plants with kettle traps, temporarily capture their pollinators [39]. With special adhesive devices, insects, *e. g.* flies are able to attach to rough or smooth surfaces. On rough surfaces, claws allow anchorage via hooking, whereas smooth adhesive pads or hairy pads enable adhesion via van-der-Waals and capillary forces on smooth surfaces [40 – 42]. Interestingly, these mechanisms fail on the plant traps and the insects lose their foothold. Such slippery plant surfaces might play an important role as templates for a transfer into technical materials for insect pest control. Poppinga *et al.* [37] revealed that in pitfall traps combinations of epidermal cell curvatures, cuticular folding, three-dimensional epicuticular wax crystals and idioblastic elements exist. The most prevalent cell shape, found in 35 species, is papillae cells with downwards leading orientation. 29 species showed two or three levels of surface sculptures. Examples of two-level sculptures are *Aristolochia pearcei* (Fig. 4a) with papillate epidermal cells covered with waxes and *Cephalotus follicularis* (Fig. 4b) with papillate epidermal cells and cuticular folds. Sliding structures with a three-level hierarchical sculpture were found in *e. g. Sarracenia leucophylla*, with papillate epidermal cells and cuticle folding with superimposed epicuticular waxes (Fig. 4c).

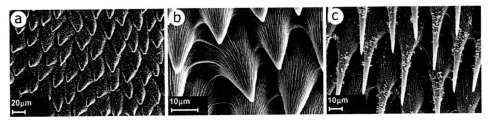

Figure 4. Scaning electron micrographes of slipery plant surfaces. In **(a)** *Aristolochia pearcei* the surface sculptures are formed by papillate epidermal cells covered with waxes. In **(b)** the surface sculptures of *Cephalotus follicularis* are formed with papillate epidermal cells and cuticular folds. In *Sarracenia leucophylla* **(c)** downward pointing, papillate epidermal cell shape care combined with two superimposed fine structures (epicuticular waxes and cuticular folds). Photos from [37].

CONCLUSIONS

In most plants superhydrophobic surfaces are formed by microstructured cells with three-dimensional superimposed waxes, or by multi-cellular hairs, which are also covered by three-dimensional waxes. Hierarchical roughness of plant surfaces leads to self-cleaning or underwater air-retaining surfaces, as shown for the lotus leaves and the water fern *Salvinia*. Anti-adhesive "sliding structures" of plants are further examples of highly functional biological interfaces.

The structures of plant surfaces and their remarkable functions introduced here demonstrate that nature provides solutions for the development of artificial functional materials. The surface structures presented here are only a small fraction of existing structures, but they might stimulate the research in biomimicry, and might help to transfer the ideas and concepts implemented in nature into technology.

REFERENCES

[1]   Bar-Cohen, Y. (2006), Biomimetics-using nature to inspire human innovation. *Bioinspiration and Biomimetics* **1**:1 – 12.
        doi: 10.1088/1748-3182/1/1/P01.

[2]   Koch, K., Bhushan, B. and Barthlott, W. (2009a) Multifunctional Surface Structures of Plants: An Inspiration for Biomimetics. *Prog. Mater. Sci.* **54**:137 – 178.
        doi: 10.1016/j.pmatsci.2008.07.003.

[3]   Jeffree, C.E., (2006) The fine structure of the plant cuticle. In *Biology of the plant cuticle*, (Eds. Riederer, M. and Müller, C.). *Blackwell Oxford UK*, pp. 11 – 125.
        doi: 10.1002/9780470988718.ch2.

[4]   Kerstiens, G., (1996a) *Plant cuticles: an integrated functional approach.* Bios Scientific Publisher, Oxford, UK.

[5]   Riederer, M. and Müller, C. (2006) *Biology of the plant cuticle.* Blackwell Oxford UK.
        doi: 10.1002/9780470988718.

[6]   Bargel, H., Neinhuis, C. (2006) Tomato (*Lycopersicon esculentum* Mill.) fruit growth and ripening as related to the biomechanical properties of fruit skin and isolated cuticle. *J. Exp. Bot.* **56**(413):1049 – 60.
        doi: 10.1093/jxb/eri098.

[7]   Barthlott, W., Neinhuis, C., Cutler D., Ditsch F., Meusel I., Theisen I. and Wilhelmi, H. (1998) Classification and Terminology of Plant Epicuticular Waxes. *Bot. J. Linn. Soc.* **126**:237 – 260.
        doi: 10.1111/j.1095-8339.1998.tb02529.x.

[8]   Ensikat, H.J., Boese, M., Mader, W., Barthlott, W., Koch, K. (2006) Crystallinity of plant epicuticular waxes: electron and X-ray diffraction studies. *Chem. Phys. Lipids* **144**(1):45 – 59.
        doi: 10.1016/j.chemphyslip.2006.06.016.

[9]   Koch, K., Bhushan B., Barthlott, W. (2010) Functional plant surfaces, smart materials. In *Handbook of Nanotechnology,* (ed. Bhushan B.) 3rd Ed. Springer, Heidelberg,
Germany. pp. 1399 – 1436.
doi: 10.1007/978-3-642-02525-9_41.

[10]   Koch, K., Ensikat, H.J. (2007) The hydrophobic coatings of plant surfaces: epicuticular wax crystals and their morphologies, crystallinity and molecular self-assembly.
Micron. **39**(7):759 – 72.
doi: 10.1016/j.micron.2007.11.010.

[11]   Koch, K., Dommisse, A., Niemietz, A., Barthlott, W., Wandelt, K. (2009c) Nanostructure of epicuticular plant waxes: Self-assembly and molecular architecture of wax
tubules. *Surf. Sci.* **603**:1961–1968.
doi: 10.1016/j.susc.2009.03.019.

[12]   Jetter, R., Kunst, L. and Samuels, A.L., (2006) Composition of plant cuticular waxes.
In: *Biology of the plant cuticle,* Riederer, M. and Müller, C. (Eds.), Annual Plant
Reviews 23, *Blackwell Oxford UK* pp. 145 – 175.

[13]   Jetter, R. and Schäffer, S., (2001) Chemical composition of the *Prunus laurocerasus*
leaf surface. Dynamic changes of the epicuticular wax film during leaf development.
Plant Phys. **126**:1725 – 1737.
doi: 10.1104/pp.126.4.1725.

[14]   Martin, C. and Glover, B.J. (2007) Functional aspects of cell patterning in aerial
epidermis. *Curr. Op. Plant Biol.* **10**:70 – 82.
doi: 10.1016/j.pbi.2006.11.004.

[15]   Barthlott, W. and Ehler, N. (1977) *Rasterelektronenmikroskopie der Epidermis-Ober-
flächen von Spermatophyten*, Tropische und Subtropische Pflanzenwelt, Akad. der
Wiss. und Lit. Mainz.

[16]   Li, W. and A. Amirfazli, A. (2008) Hierarchical structures for natural superhydrophobic surfaces. *Soft Matter* **4**:462 – 466.
doi: 10.1039/b715731b.

[17]   Forbes, P. (2008) Self-cleaning materials. *Sci. Am.* **299**:88 – 96.
doi: 10.1038/scientificamerican0808-88.

[18]   Barthlott, W., Schimmel, T., Wiersch, S., Koch, K., Brede, M., Barczewski, M.,
Walheim, S., Weis, A., Kaltenmaier, A., Leder, A., Bohn, H.F. (2010) The Salvinia
paradox: superhydrophobic surfaces with hydrophilic pins for air-retention under
water. *Adv. Mater.* **22**:2325–2328.
doi: 10.1002/adma.201090075.

[19]   Nosonovsky, M. and Bhushan, B. (2008) *Multiscale Dissipative Mechanisms and Hierarchical Surfaces: Friction, Superhydrophobicity, and Biomimetics.* Springer-Verlag, Heidelberg, Germany

[20]   Koch, K., Bhushan, B. and Barthlott, W. (2008) Diversity of Structure, Morphology and Wetting of Plant Surfaces. *Soft Matter* **4**:1943 – 1963.
       doi: 10.1039/b804854a.

[21]   Barthlott, W. and Neinhuis C. (1997) Purity of the sacred lotus, or escape from contamination in biological surfaces. *Planta,* **202**:1 – 8.
       doi: 10.1007/s004250050096.

[22]   Patankar, N.A. (2004) Mimicking the Lotus Effect: Influence of Double Roughness Structures and Slender Pillars. *Langmuir* **20**:8209 – 8213.
       doi: 10.1021/la048629t.

[23]   Neinhuis, C. and Barthlott, W. (1997) Characterization and Distribution of Water-repellent, Self-cleaning Plant Surfaces. *Ann. Bot.* **79**:667 – 677.
       doi: 10.1006/anbo.1997.0400.

[24]   Roach, P., Shirtcliffe, N.J., Newton, M.I. (2008), Progress in superhydrophobic surface Development. *Soft Matter* **4**:224 – 240.
       doi: 10.1039/b712575p.

[25]   Shirtcliffe, N.J., McHale, G., Newton, M.I., Chabrol, G., Perry, C.C. (2004) Dual-scale roughness produces unusually water-repellent surfaces. *Adv. Mater.* **16**:1929 – 1932.
       doi: 10.1002/adma.200400315.

[26]   Fürstner, R., Barthlott, W., Neinhuis, C. and Walzel, P. (2005) Wetting and self-cleaning properties of artificial superhydrophobic surfaces. *Langmuir* **21**:956 – 961.
       doi: 10.1021/la0401011.

[27]   Nosonovsky, M. and Bhushan, B., (2007) Multiscale friction mechanisms and hierarchical surfaces in nano and bio-tribology. *Mater. Sci. Eng. R* **58**:162 – 193.
       doi: 10.1016/j.mser.2007.09.001.

[28]   Bhushan, B., Jung, Y.C., Koch, K. (2009) Self-cleaning efficiency of Artificial Super-hydrophobic Surfaces. *Langmuir* **25**:3240 – 3248.
       doi: 10.1021/la803860d.

[29]   Wenzel, R.N. (1936) Resistance of solid surfaces to wetting by water. *Ind. Eng. Chem. Res.* **28**:988.
       doi: 10.1021/ie50320a024.

[30]   Cassie, A.B.D. and Baxter, S. (1944) Wettability of porous surfaces. *Trans. Faraday Soc.* **40**:546.
         doi: 10.1039/tf9444000546.

[31]   Herminghaus, S. (2000) Roughness-Induced Non-Wetting. *Europhys. Lett.* **52**:165 – 170.
         doi: 10.1209/epl/i2000-00418-8.

[32]   Bhushan, B., Jung, Y.C., Koch, K. (2009) Micro-, Nano-, and Hierarchical Structures for Superhydrophobicity, Self-Cleaning and Low Adhesion. *Philos. Trans. R. Soc., A, Biomimetics II: fabrication and applications* **367**:1631 – 1672.

[33]   Ensikat, H.J., Schulte, A.J., Koch, K., Barthlott, W. (2009) Droplets on Superhydrophobic Surfaces: Visualization of the Contact Area by Cryo-Scanning Electron Microscopy. *Langmuir* **25**:13077 – 13083.
         doi: 10.1021/la9017536.

[34]   Koch, K., Bohn, H.F. and Barthlott, W. (2009b) Hierarchical sculpturing of plant surfaces and Superhydrophobicity. *Langmuir special issue "Wetting and Superhydrophobicity"* **25**:14116 – 14120.

[35]   Eyring, V., Köhler, H.W., Aardenne, J. van, Lauer, A. (2005) Emissions from international shipping: 1. The last 50 years. *J. Geophys. Res., C: Oceans Atmos.* **110**:17305.
         doi: 10.1029/2004JD005619.

[36]   Fukuda, K., Tokunaga, J., Nobunaga, T., Nakatani, T., Iwasaki, T. and Kunitake, Y. (2000) Frictional drag reduction with air lubricant over a super-water-repellent surface. *J. Mar. Sci. Tech.* **5**:123 – 130.
         doi: 10.1007/s007730070009.

[37]   Poppinga, S., Koch, K., Bohn, H.F., Barthlott W. (2010) Comparative and functional morphology of hierarchically structured anti-adhesive plant surfaces. *Funct. Plant Biol.* **37**(10):952 – 961.
         doi: 10.1071/FP10061.

[38]   Juniper, B.E., Robins, R.J., Joel, D.M. (1989) The carnivorous plants. Academic Press, London.

[39]   Vogel, S., Martens, J. (2000) A survey of the function of the lethal kettle traps of Arisaema (Araceae), with records of pollinating fungus gnats from Nepal. *Bot. J. Linn. Soc.* **133**:61 – 100.
         doi: 10.1111/j.1095-8339.2000.tb01537.x.

[40] Beutel, R.G. and Gorb, S.N. (2001) Ultrastructure of attachment specializations of hexapods (Arthropoda): evolutionary patterns inferred from a revised ordinal phylogeny. *J. Zool. System. Evol. Res.* **39**:177–207.
      doi: 10.1046/j.1439-0469.2001.00155.x.

[41] Gorb, S.N. (2008a) Smooth attachment devices in insects: functional morphology and biomechanics. *Adv. Insect Physiol.* **34**:81–115.
      doi: 10.1016/S0065-2806(07)34002-2.

[42] Gorb, S.N. (2008b) Biological attachment devices: exploring nature's diversity for biomimetics. *Philos. Trans. R. Soc., A* **366**:1557–1574.
      doi: 10.1098/rsta.2007.2172.

Towards Electron Beam Induced Deposition Improvements for Nanotechnology

Johannes J.L. Mulders[1,*] and Aurelien Botman[2]

[1]FEI Electron Optics, Achtseweg Noord 5, 5600 KA Eindhoven, The Netherlands

[2]FEI Company, 5350 NE Dawson Creek Drive, Hillsboro, OR 97124, U.S.A.

E-Mail: *jjm@fei.com

Received: 25th August 2010 / Published: 13th June 2011

Abstract

Electron beam induced deposition can be applied as a direct write technique for the creation of 3 dimensional nano-scale structures. The technique does not require lift-off and mask based process steps and therefore has potential for application in rapid prototyping for nano-technology, with a high degree of flexibility. The material quality and microstructure of the deposition however, is usually somewhat different to that of the pure material and hence the obtained properties of the nano-depositions do not reflect the bulk properties of the desired material. Current research is focused on improvements for the deposition processes with the aim to improve the purity of the deposition to such an extent that the local characteristic aimed for (such as conductivity) is good enough to attain the required local functionality. This paper reports on the current status and on some of the methods that have been applied to improve the deposition process.

Introduction

Electron beam induced deposition (EBID) is a technique that can be used for the direct creation of nanoscale structures in 3 dimensions. The base technique includes a scanning electron microscope (SEM) equipped with a gas supply line providing precursor molecules released in the vacuum chamber of the SEM and close to the substrate. These molecules will temporarily adhere to the surface and within that residence time they may interact with electrons supplied by the electron beam (Figure 1 shows a schematic representation of the process).

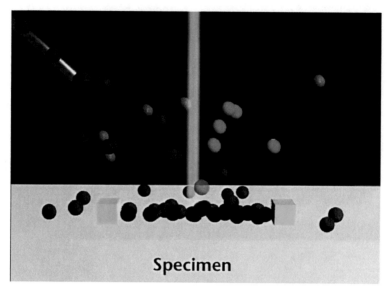

Figure 1. Schematic representation of the EBID process: Precursor molecules (red) are decomposed by interaction with the electron beam into a volatile (green) and non-volatile (blue) part. By beam patterning a conductive pad is created between the two contact pads (yellow).

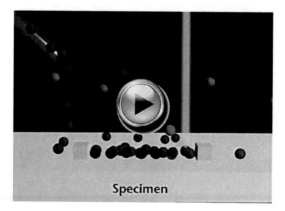

Movie. Principle of EBID process

This interaction results in a decomposition of the precursor molecule into a volatile part such as a CO or a CH_4 group (that is pumped away by the vacuum system) and a non-volatile part such as the metal of interest that forms a very local deposition on the substrate. Typically the local flux of molecules in the order of 10^{16} mol/cm^2.s, while beam currents are in the range 50 pA to 1 nA. It is generally assumed that the secondary electrons with an energy < 50 eV, emitted from the substrate, are the main contributors to the decomposition process. An example of a pattern deposited using W(CO)$_6$ as a precursor is shown in Figure 2.

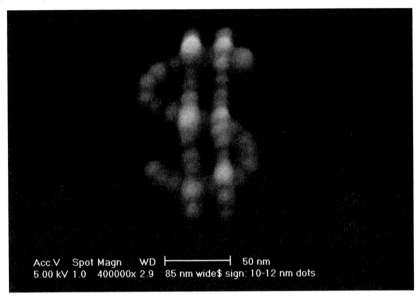

Acc.V Spot Magn   WD ├─────────────┤ 50 nm
5.00 kV 1.0   400000x 2.9   85 nm wide$ sign: 10-12 nm dots

Figure 2. Example of an EBID created dollar symbol, showing the direct nano-patterning capability. The image shows nano-dots of 10 – 12 nm made by the decomposition of W(CO)$_6$ by the electron beam exposure at 3 seconds per point and a total production time of 150 seconds.

Typically for the creation of metallic structures the precursor is an organo-metallic chemical, the vapour of which is used as a gas supply to the system. Many of these chemicals are used in CVD based processes, where decomposition is driven by collisions with the substrate at high temperature (thermal decomposition). Examples of precursor materials that are regularly used for EBID by many users across the world are listed in Table 1.

Table 1. Overview of commonly used precursors for electron beam induced deposition

Precursor	Short name / formula	For deposition of
methyl cyclo-pentadienyl platinum trimethyl	MeCpPtMe$_3$	Platinum
tungsten hexacarbonyl	W(CO)$_6$	Tungsten
naphthalene	C$_{10}$H$_8$	Carbon
cobalt octacarbonyl	Co$_2$(CO)$_8$	Cobalt
tetra ethyl ortho silicate	TEOS	SiO$_x$
dimethyl gold acetyl acetonate	Me$_2$Auacac	Gold

Applications for these materials range from contacting nanotubes and nanowires (conductivity, work function), seeding for nanowire growth, electrical isolation, optical and chemical activity and nanoscale magnetic structures such as for domain wall movement or tips for Magnetic Force Microscopy (MFM). It should be noted that the precursors listed in Table 1 all contain carbon atoms in some form (methyl, carbonyl groups) and ideally the decomposition reaction should only decompose into non-volatile (metal) and volatile products such as CO. For example, ideally the creation of nanoscale tungsten depositions should be close to the following reaction:

$$W(CO)_6 + electrons \quad \rightarrow \quad W \downarrow \quad + \quad 6\ CO \uparrow + electrons$$

In practice however, the decomposition is not 100% effective and hence there is a co-deposition of both carbon and oxygen, together with the tungsten. Typically for the process described above, the elements present in the deposition are: W 37 at%, O 27 at% and C 36 at%. For this reason the local conductivity of the structure will deviate substantially from the value for bulk tungsten. This applies to any relevant characteristic of the deposition: the local nanoscale property is generally quite different from the known bulk property due to the presence of undesired elements. Therefore current research focuses on a better understanding of the fundamental processes [1, 2], optimization of the basic parameters [3], additional *ex situ* purification techniques [4] and the application of other, carbon free precursors [5]. In this paper the purification of platinum depositions is discussed as well as the use of alternative (non-standard) precursors for platinum and gold and an outlook into further improvements and new materials.

CURRENT STATUS OF THE TECHNOLOGY

In literature it is found that the number of publications on the subject of EBID is increasing and a wide range of materials and processes have been tried [6]. The basic processes of the technique, such as the individual cross-section for the reaction of energetic electrons with adsorbed molecules is in its infancy and is sometimes studied by looking at reaction products (RGA) and X-ray photo-electron spectroscopy (XPS) *in situ*, when the adsorbed precursor is irradiated with an electron broad beam. These techniques have a sensitivity that is not adequate for the small amount of material involved when performing EBID at nanoscale, but for a broad and low energy electron beam producing a thin film, they are well suited. From these measurements it can be derived how the material decomposes and how reaction products are formed. Also the actual growth dynamics have now been simulated by computer models [7] and the basic growth parameters are increasingly better understood. Overall the research focuses on understanding the basic reaction mechanisms, both from a chemical point of view and from a dynamic point of view. In addition many methods are investigated to get a better purity with minimal or no changes of the 3 dimensional shape.

One thing which is very clear from literature is that the total parameter space is very large and, more importantly, not all parameters are listed when experimental results are presented. As a consequence the reproducibility of results is poor. The main reason for this is a non-consistent approach from user to user and from tool to tool, indicating the presence of one or more process parameters that have not yet been recognized. As an example of one of these parameters, it has been observed through recent experiments, that the vacuum and cleanliness conditions are quite crucial for the results. This is because the processes involved include real surface reactions with low amounts of material and hence any molecules that may be present can add to the process. Vacuum conditions are generally not reported in literature. Even the exact procedure used by the SEM operator may come to influence the process, due to a lack of awareness of all the critical factors involved in good experimental reproducibility. When performing EBID experiments it is highly recommended for an operator to apply the following conditions:

1) Related to the chamber vacuum:

   a) waiting until the base vacuum of the system is at least $<4 \times 10^{-6}$ mbar;

   b) minimizing the amount of residual water vapour by using a cold finger or an overnight pump out;

   c) controlling residual hydrocarbons by working in very clean conditions and by applying a regular plasma clean of the chamber;

   d) regularly confirming that the Penning gauge readout is correct: EBID processes tend to increase the contamination of the interior of the Penning indicating the pressure to be better than in reality.

2) Related to sample mounting:

   a) not using double-sided adhesive carbon tabs;

   b) not using silver paint or carbon paint;

   c) being suspicious to any part which might produce additional out-gassing, such as plastics or other non-vacuum compatible materials;

   d) using a clamping mechanism to fix the sample to the holder instead.

3) Related to sample holders:

   a) not using gel packs;

   b) not using membrane packs;

   c) using a single, clean container for each individual sample (Teflon or glass);

   d) placing a clear marker (*e.g.* using a Focused Ion Beam) on the sample for later locating the small deposited material (for example for AFM measurements).

Ideally the sample surface should be well defined (material type and crystal orientation) and the vacuum should be quantified with more data than just the pressure readout: a mass spectrometer spectrum can provide more informative and detailed data about the actual vacuum present during the experiment.

RECENT PROGRESS ON PURIFICATION OF PLATINUM

For platinum the most widely used precursor is MeCpPtMe$_3$, mostly used for contacting nanotubes and nanowires. The challenge in improving the EBID process for platinum starts with an analysis of the main system parameters that have an influence on the actual composition. These refer to beam energy and beam current. The relation between the amount of platinum as determined by low kV EDX, as a function of beam current and energy is shown in Figure 3. As beam current and beam energy together define the power of the beam it is also interesting to plot the power against the platinum content, as shown in Figure 4.

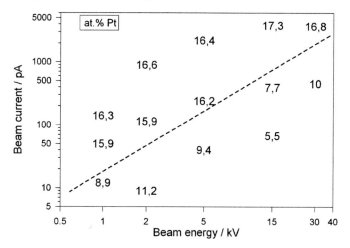

Figure 3. The amount of Pt given in numbers as atomic% (at%) as a function of the applied beam energy and beam current. The dashed line splits the parameter space into two regions: ~16 at% and < 16 at% [8].

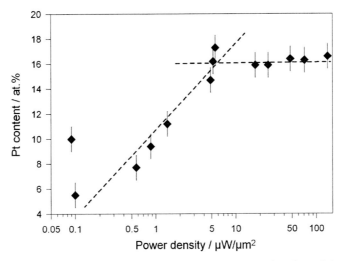

Figure 4. Same data as in Figure 3, but now expressed as a function of the beam power density, defined here as the beam current × beam energy/projected electron interaction area on the substrate surface as derived from Monte Carlo simulations [8].

From both figures it can be concluded that:

1. The higher the power density, the higher the amount of platinum. It is suggested that additional heat generation during the deposition process may deliver extra energy and hence enhance the decomposition, as is also suggested for decomposition of cobalt [9].

2. There appears to be an upper limit of purity around 16 at% Pt. This suggests a decomposition where 4 CH_4 groups are removed from the MeCpPtMe$_3$, leaving the remaining pentadienyl ring bound to the platinum. The reaction could then be proposed as follows:

$$C_9H_{16}Pt + \text{electrons} \rightarrow 4\,CH_4 \uparrow + C_5Pt \downarrow + \text{electrons}$$

resulting in 1 Pt atom out of 6 per deposited molecule, i.e. 16 at%.

A further improvement could be obtained by post treatment of the deposition:

1. Annealing in oxygen [4]: this results in 70 at% Pt when applied in the temperature range 250 – 500°C and is generally attributed to a burn-off of the carbon content and the release of CO or CO_2.

2. Annealing in a hydrogen radical environment [10]: this results in an improvement of the platinum content to 35 at% in the temperature interval 150 – 180°C. At higher temperatures no further changes have been observed, but it is remarkable that the improvement starts at a lower temperature, most likely by the generation of CH_4.

Although the improvement in the relative amount of Pt is clear it should be noted that with these treatments the substrate will be subjected to a relative high temperature (not always desirable for the sample as a whole). In consequence the shape of the deposited material can change quite a lot, due to the violent and non-symmetric removal of material.

Another way to create Pt depositions is to use a non-organic precursor $Pt(PF_3)_4$ [11]. This precursor contains PF_3 ligands and may therefore produce some fluorine-containing molecules during decomposition, such as F_2 or HF. The sample should be resistant to these etching molecules. Quantitative measurements of small-sized depositions using the standard EDX technique is not straightforward because the $PtM\alpha_1$ X-ray line at 2.05 keV is very close to the $PK\alpha$ line at 2.013 keV and hence these lines cannot be very well distinguished by the EDX system, which typically has an energy resolution of around 120 eV. In comparative analysis one can either use the combined X-ray peak or use TEM or WDX at higher energy for analysis of the PtL lines.

Similar to the $MeCpPtMe_3$ precursor, post processing can be applied to improve the purity. It should be noted that for platinum the main application is focused on electrical conductivity and hence the specific resistivity ($\mu\Omega.cm$) is the parameter of interest and this can be measured by using a 4 point probe test structure as shown in Figure 5. In this typical structure the resistivity can be measured without influence of the contact resistance and by measuring the cross section the resistivity can be calculated.

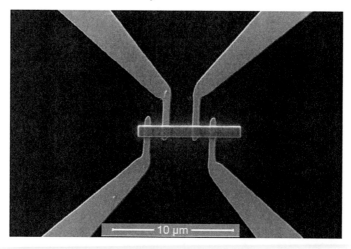

Figure 5. A 4 point probe test structure used to measure the resistance across an EBID strip. By measuring the height of the strip the cross-section can be determined and hence the material resistivity. The end of the strip can also be used for compositional measurement using EDX without affecting the portion of the strip used for resistivity determination.

Although the composition and purity of the deposition plays a dominant role in the final electrical property of the material, it should also be noted that the local fine structure (grain size) and crystalline distribution also contribute to the final result. These can be determined by TEM based analysis and often reveal small pure Pt grains embedded in a matrix containing the other elements such as C for MeCpPtMe$_3$ or P and O for Pt(PF$_3$)$_4$. All improvements that have been obtained for platinum depositions can best be summarized by looking at the progress in the decrease of the deposited material resistivity, as shown in Figure 6. From this figure it is obvious that much progress has been made by the various methods and post processing and that the conductivity is now so close to the value for bulk Pt, that the EBID process may become useful for the application it was intended for (such as nano-contacts).

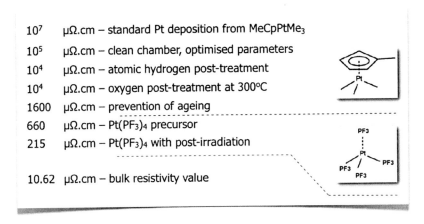

10^7	µΩ.cm – standard Pt deposition from MeCpPtMe$_3$
10^5	µΩ.cm – clean chamber, optimised parameters
10^4	µΩ.cm – atomic hydrogen post-treatment
10^4	µΩ.cm – oxygen post-treatment at 300°C
1600	µΩ.cm – prevention of ageing
660	µΩ.cm – Pt(PF$_3$)$_4$ precursor
215	µΩ.cm – Pt(PF$_3$)$_4$ with post-irradiation
10.62	µΩ.cm – bulk resistivity value

Figure 6. Substantial improvements in the specific resistivity for EBID Pt, using a variety of purification methods and/or a non-organic precursor [8].

RECENT PROGRESS ON PURIFICATION OF GOLD

The interest in gold for nano-depositions is driven by its use as a nanoseed for growing of III/V semiconductor nanowires, as an optical active medium for plasmonic structures and as a marker and binding site for biological structures. In addition it also can be used for creating electrical contacts. The most commonly applied precursor is Me$_2$Au(acac) and this precursor can indeed be used for EBID but the amount of Au is only 8 – 10 at% with the remainder being mainly C and a little O. Also for this precursor a post anneal step in a nitrogen environment is possible to improve the purity, as shown in Table 2.

Table 2. Composition (at%) of EBID from Me₂Au(acac). Depositions made at room temperature. Anneal steps at 250 and 400°C in a nitrogen environment. The anneal steps were carried out at ICFO, Barcelona Spain.

Element	Sample 1 no anneal	Sample 2 250°C -3 hr	Sample 3 400°C -3 hr
C	79.1	78.9	54
O	11.4	11.6	20.4
Au	9.5	9.5	25.6
	100	100	100

More recently annealing was also applied in an oxygen environment, which is assumed to be more active for assisting the burn-off of the carbon [10] than a relatively inert environment of nitrogen would be. This is confirmed by the result shown in Figure 7. The carbon content at 400 °C drops to around 22 at% for anneal in oxygen, while for anneal in nitrogen the carbon content reduces to around 54 at%.

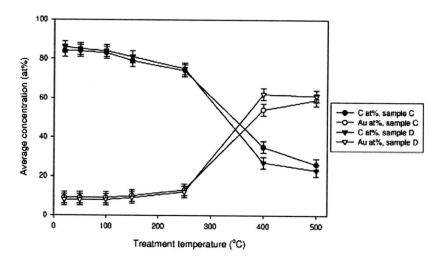

Figure 7. EBID using Me₂Au(acac). The amount of gold (at%) as a function of the anneal temperature using an oxygen environment for 10 minutes. The curve shows a saturation level around 60 at%. This figure was first published in [10] and is reproduced with permission.

Although the results from both anneal steps are encouraging, it should be noted that the shape of the nano-deposit suffers. At even higher temperatures the Au may also coagulate and form larger isolated clusters on the substrate.

Application of a non-organic precursor AuClPF₃ is reported in literature [12] and this produces very high purity gold. However this precursor is not commercially available and not very stable (it decomposes spontaneously quite easily at room temperature). In addition it releases fluorine during the deposition process which may attack the substrate, sample and/or the system. Therefore, recently a new precursor was tested: Au(CO)Cl. This material has been known since 1925 and is readily available. An actual deposition with this precursor is shown in Figure 8. A comparison of the purity of the gold deposition is shown in the spectra in Figure 9 for $Me_2Au(acac)$ and Au(CO)Cl respectively. The quantitative purity is better than 92 at% [13], the yield is 6.5×10^{-4} $\mu m^3/nC$ for 5 kV, 100 pA and in the gas limited regime. The deposition shows a grain structure. The yield is relatively low and this is due to the low gas flux. However, the gas flux cannot be increased by precursor heating due to enhanced spontaneous chemical decomposition.

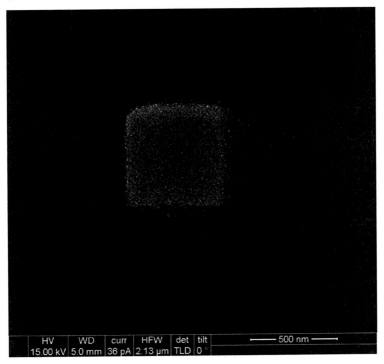

HV	WD	curr	HFW	det	tilt	——— 500 nm ———
15.00 kV	5.0 mm	36 pA	2.13 μm	TLD	0 °	

Figure 8. Deposition of a 500 × 500 nm pad at 15 kV, 25pA with 15 minutes total time.

The yield is relatively low, but the purity is very good.

190

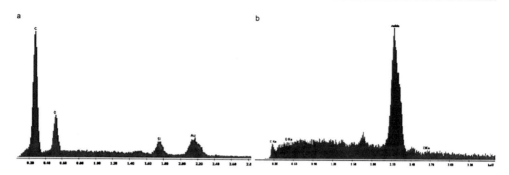

Figure 9. (a) EDX spectrum comparison of EBID with Me2Au(acac) giving ~8 at% Au; **(b)** EBID with Au(CO)Cl giving >92 at% Au.

The precursor Au(CO)Cl allows the production of very pure gold structures on substrates that are compatible with the release of some chlorine, such as SiO_2 and Si_3N_4. It also allows the creation of very small nanoscale dots as shown in Figure 10. It shows that the material has potential for application in plasmonic structures for the manipulation of light, using gold antennas at sub-wavelength dimensions.

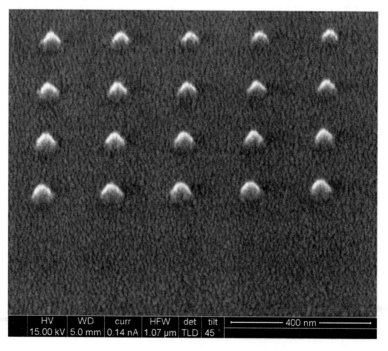

Figure 10. Part of a 15 × 15 dot array. Point depositions at 15 kV, 25 pA. Diameter 50 nm, height is around 40 nm. Each dot is high purity gold deposited from Au(CO)Cl. Production time of the total array is 5 minutes.

OUTLOOK AND DISCUSSION

As illustrated above, recent developments in the search for higher purity of the deposited material have resulted in substantially higher atomic content of the desired metal, both for Pt and for Au. Therefore application of these materials is now within reach of the nanotechnology scientist. Although the local property of interest is not yet at the level of the bulk material, the property gets very close to this and becomes practical for use. A second trend that is ongoing is the broadening of the materials range including magnetic materials such as Fe, Co and Ni. Co has been widely studied and has already been shown to exhibit ferromagnetic behaviour [3] and for Fe similar results are emerging. Other materials of interest might be Pd [14], and in view of the lessons learned for Pt these results can be expected to improve in the near future.

Finally it is interesting to observe the use of EBID together with other techniques where the advantages of both can be combined. One such opportunity is the combination of atomic layer deposition (ALD) with seeds made by EBID [15], both using the same MeCpPtMe$_3$ precursor, but now resulting in 100% Pt which is driven by the ALD process. It has been shown that the structural behaviour of EBID-seeded ALD growth provides the same purity and crystal structure (texture) as films grown by ALD alone. The combination of the two techniques also allows a higher total process speed, if the area to cover is relatively large: this is due to the fact that ALD is a parallel growth technique and the required EBID seeds can be very thin, on the order of 1 nm only.

REFERENCES

[1]   Smith, D.A., Fowlkes, J.D., Rack, P. D. (2007) *Nanotechnology* **18**:265308. doi: 10.1088/0957-4484/18/26/265308.

[2]   Wnuk, J.D., Gorham, J.M., Rosenberg, S.G., van Dorp, W.F., Madey, T.E., Hagen, C.W. and Fairbrother, D.H. (2010) *J. Vac. Sci. Technol.* **B** (in press).

[3]   Fernandez-Pacheco, A., De Teresa, J.M., Cordoba, R. and Ibarra, M.R. (2009) *J. Phys. D, Appl. Phys.* **42**:055005.

[4]   Botman, A., Mulders, J.J.L., Weemaes, R. and Mentink, S. (2006) *Nanotechnology* **17**: 3779 – 3785. doi: 10.1088/0957-4484/17/15/028.

[5]   Barry, J.D., Ervin, M., Molstad, J., Wickenden, A., Britlinger, T., Hoffmann, P. and Melngailis, J. (2006) *J. Vac. Sci. Technol.* B **24** (6):3165 – 68. doi: 10.1116/1.2395962.

[6]   Botman, A., Mulders, J.J.L. and Hagen, C.W. (2009) *Nanotechnology* **20**:372001.
      doi: 10.1088/0957-4484/20/37/372001.

[7]   Rack, P.D., Fowlkes, J.D., Randolph, S.J., Smith, D.A. (2006) *First International Workshop on Electron Beam Induced deposition, Delft, The Netherlands,* 16 – 17.

[8]   Botman, A. (2009) *Thesis* TU Delft, ISBN 978 – 90 – 5335 – 240 – 3.

[9]   Córdoba, R., Sesé, J., De Teresa, J.M., Ibarra, M.R. (2009) *Microelectronic Engineering* **87**(5 – 8):1550 – 1553.

[10] Botman, A., Hesselberth, M., and Mulders, J.J.L. (2008) *Microelectronic Engineering* **85**:1139 – 1142.
      doi: 10.1016/j.mee.2007.12.036.

[11] Barry, J.D., Ervin, M., Molstad, J., Wickenden, A., Britlinger, T., Hoffmann, P. and Melngailis, J. (2006) *J. Vac. Sci. Technol.* B, **24** (6):3165 – 68.
      doi: 10.1116/1.2395962.

[12] Hoffmann, P., Utke, I., Cicoira, F., Dwir, B., Leifer, K., Kapon, E. and Doppelt, P. (2000) *Mat. Res. Soc. Symp. Proc.* **624**:171.

[13] Mulders, J.J.L., Botman, A. (2009) *Procedings Microscopy and MicroAnalysis* Richmond USA, 1126.

[14] Spoddig, D., Schindler, K., Roediger, P., Barzola-Quiquia, J., Fritsch, K., Mulders, J. and Esquinazi, P. (2007) *Nanotechnology* **18**:495202
      doi: 10.1088/0957-4484/18/49/495202.

[15] Mackus, A.J.M., Mulders, J.J.L., van de Sanden, M.C.M and Kessels, W.M. (2010) *Journal of Applied Physics* **107**:116102 – 2010.
      doi: 10.1088/0957-4484/18/49/495202.

Focused Electron Beam Induced Deposition – Principles and Applications

Michael Huth

Physikalisches Institut, Goethe-Universität, Max-von-Laue-Str. 1,
60438 Frankfurt am Main, Germany

E-Mail: michael.huth@physik.uni-frankfurt.de

Received: 27th August 2010 / Published: 13th June 2011

Abstract

Focused electron beam induced deposition (FEBID) is a direct beam writing technique for nano- and micro-structures. By proper selection of the precursor gas, which is dissociated in the focus of the electron beam, different functionalities of the resulting deposits can be obtained. This contribution discusses nano-granular FEBID materials. Quite generally, nano-granular metals can be considered as tunable model systems for studying the interplay of electronic correlation effects, quantum size effects and disorder. After the introduction into the FEBID process a brief overview of the different electronic transport regimes in nano-granular metals is given. Recent experimental results on electron irradiation effects on the transport properties are presented. These results indicate a new methodology for highly miniaturized strain sensor element fabrication based on the specific electronic properties of nano-granular FEBID structures.

Introduction

Anyone who has used a scanning electron microscope (SEM) will have noticed that the area over which the electron beam is rastered for image acquisition tends to become covered with a thin film of a material which provides a rather low secondary electron yield, i. e. appears dark. This thin film, of a few nm thickness, is formed by the non-volatile electron beam induced dissociation products of hydrocarbons adsorbed on the specimen surface. The hydrocarbons themselves are part of the typical residual gas atmosphere of the SEM's vacuum chamber. Already in 1976 this electron-beam induced dissociation phenomenon has been used for demonstrating the nano-patterning capabilities of focused electron beam induced deposition (FEBID) down to the sub-100 nm scale [1]. In the 1980s other gases were deliberately introduced in SEMs to study the results of dissociation processes with a view to obtaining deposits which might be able to provide certain functionalities, such as high metallic conductivity [2 – 4]. In the following years numerous precursor gases were systematically tested and various 2D and 3D structures were fabricated. Nevertheless, the beginning of a rather strong increase of activity in this field dates only back eight years. Since about 2002 the average number of publications and citations in the field of FEBID has increased by a factor of about 15 [5]. This can be attributed to the availability of high-resolution SEMs, often in combination with an ion-optical column for focused ion beam (FIB) processing, with commercial precursor gas injection systems. In parallel to this technological advancement FEBID, in conjunction with focused electron beam induced etching, is now routinely used in high-end tools for photolithographic mask repair in the semiconductor industry [6].

In many instances and for a large variety of precursors the structures obtained by the FEBID process are nano-granular, i.e. they are formed by a composite consisting of metallic nano-crystallites embedded in an insulating carbonaceous matrix. This has important consequences. On the one hand, the nano-granular structure leads to a significant increase of the resistivity as compared to that of the pure metal. Consequently, strong efforts are made to improve on the metal content of FEBID structures with the ultimate goal of reaching 100% pure metal deposits for a wide range of applications in mask repair and circuit editing. On the other hand, the nano-granularity influences the elastic properties of FEBID structures. Recent research has shown examples of very large hardness, approaching that of diamond, as well as rubber-like behaviour in FEBID nano-pillars depending on the precursor and process parameters [7]. And finally, the nano-granularity leads to a wealth of exciting phenomena in the electronic properties of FEBID structures. Nano-granular materials provide a model system with tunable parameters suitable for studying the interplay of electron correlations, dimensionality, and the effects of mesoscopic disorder on the electronic properties; for a recent theoretical review see [8]. From the experimental point of view the study of the electrical transport properties of nano-granular FEBID structures with particular emphasis on correlation effects has begun only recently [9, 10]. Also, it has been recognized that nano-granular materials hold some promise for strain-sensing applications [11]. This will be the topic in the last part of this manuscript which will show some very recent results of the strain-resistance effect in FEBID structures.

THE FEBID PROCESS

Figure 1 shows a schematic representation of the FEBID process. Precursor molecules, supplied close to the focal point of the electron beam by a gas injection system, are dissociated by the primary electrons, backscattered electrons and secondaries. The primary electron beam is rastered over the substrate surface following a predefined pattern. Relevant process parameters for this raster process are the distance between successive dwell points of the electron beam (pitch) and the time period over which the electron beam is held at each dwell point (dwell time). Typical pitches vary between 10 to 100 nm. Dwell times vary much more strongly. Depending on the precursor and substrate material used, as well as the desired sample composition and targeted growth regime, the dwell time may a as short as 50 ns but can also be as long as 100 ms. A detailed recent review concerning FEBID and related techniques can be found in [12].

In the FEBID process the reaction of the electron beam with the precursor molecules adsorbed on the surface follows a second order kinetics, i.e. the reaction rate is proportional to both, the surface density of adsorbed molecules and the flux density of the electrons. From this proportionality one can conclude that all possible intermediate reactions leading to the final dissociation product (deposit and volatile components) have time scales which are short when compared to the time between two successive electron impact events. It is these intermediate reactions and processes of FEBID which are not yet investigated in sufficient detail.

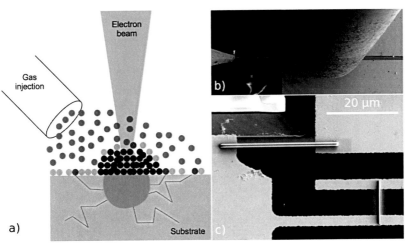

Figure 1. (a) Schematic representation of the FEBID process. The adsorbed precursor molecules (orange discs) are dissociated by electron impact (red discs) and a permanent deposit (blue discs) is formed in the focal area of the electron beam. The green lines indicate exemplary trajectories of electrons leaving the excitation volume. **(b)** SEM image of a cantilever structure with pre-defined contact lines. The gas injection capillary is visible in the upper right. On the left the W tip of a nano-manipulator is touching the cantilever. **(c)** Pt-based sensor element between contact pads (left-to-right structure) and reference element between contact pads (top-to-bottom structure) prepared by FEBID using the precursor MeCpPt(Me)$_3$.

If provided with all necessary input parameters, such as the energy-dependent dissociation cross sections and the diffusion constants for adsorbed precursor molecules, to name just two, it is possible to give a semi-quantitative account of the growth rate in FEBID on the basis of rate equation descriptions. One important ingredient in these calculations is the spatial distribution of electrons created within the Gaussian beam profile of the focussed electron beam. This distribution can be obtained from Monte Carlo simulations.

On the microscopic level there are numerous interaction mechanisms during electron impact on the precursor molecules, such as dissociation (*e. g.* by dissociated electron attachment), stimulated desorption, polymerization, and sputtering. For each of these processes an energy-dependent cross section has to be derived in order to ultimately gain more control over such aspects as purity, lateral resolution and deposition rate.

There is no theory yet that treats the FEBID process as a multi-scale problem, including microscopic and mesoscopic length scales and time scales from ultrafast (non-equilibrium processes occurring within femto seconds) to relatively slow (growth and relaxation processes requiring nanoseconds or even microseconds). First steps into tackling this multi-scale problem are currently being undertaken in the research collaboration NanoBiC funded by the Beilstein-Institut [13].

ELECTRONIC PROPERTIES OF NANO-GRANULAR METALS

At large, structures prepared by FEBID fall into the class of disordered electronic materials in which disorder exists in varying degree, depending on the process parameters and the used precursor, ranging from a few impurities in an otherwise well-ordered polycrystalline host, to the strongly disordered limit of amorphous materials. In between these extremes the material can have the microstructure of a nano-granular system and is formed of reasonably well ordered nano-crystallites embedded into a carbon-rich dielectric matrix. For those FEBID structures which fall into the weak disorder limit electronic transport can be described by the scattering of Bloch waves by impurities. The theoretical framework for calculating the transport coefficients is the Boltzmann equation for the quasi-particles. For nano-granular and, naturally so, for amorphous materials it is not possible to use this conceptual framework. Disorder must be included in the theoretical analysis right from the beginning. This implies that two additional aspects must be taken into account. The first aspect is Anderson localization, which is related to the (spatial) structure of the wave function for a single electron in the presence of a random potential. The second aspect deals with interactions between electrons in the presence of this same random potential. Electron propagation for highly disordered materials is diffusive which leads to substantial modifications to the view derived from Landau's Fermi liquid theory.

From most of the electronic transport experiments on FEBID structures which can be found in the literature it becomes readily apparent that such Anderson localization effects, which are well understood in the weak-disorder limit for uncorrelated electrons, are by far dominated by much stronger effects due to the interplay of disorder and interactions. For the latter, there is no complete theoretical framework available. In this section the focus will therefore be on some new developments in the field of disordered electronic materials, and in particular, on granular electronic systems to which most of the FEBID structures can be assigned to. The compilation of recent theoretical results is not specific to materials prepared by FEBID but covers certain aspects of granular electronic metals in general. For an in-depth study of electronic transport in granular electronic systems, for which as yet no textbooks are available, the reader is referred to the recent review by Beloborodov and collaborators [8] and references therein.

Granular metals constitute one-, two- or three-dimensional arrays of (mesoscopically different) metallic particles – or grains – which are subject to an inter-granular electron coupling due to a finite tunneling probability. The arrangement of the particles, with a typical size range from a few nm to 100 nm, can be regular or irregular. For FEBID structures the tunnel coupling is provided by the carbon-rich matrix. The matrix may also contain individual metal atoms or few-atom clusters. It represents itself a disordered electronic system and can give rise to additional conductance channels due to activated transport between localized states in the matrix. This has to be taken into account for FEBID structures with a very small volume fraction of metallic particles. In most instances it can be neglected [14].

Depending on the zero-temperature limit of the electrical conductivity σ one discriminates metallic samples, showing $\sigma(T=0) > 0$, and insulating samples, for which $\sigma(T=0) = 0$. The effects of disorder in the grain positions and in the strength of the tunnel coupling are less important for metallic samples which are characterized by strong inter-granular coupling. For low tunnel coupling, i.e. for insulating samples, the effects of irregularities become crucial and have a direct influence on the temperature dependence of the conductivity. As a consequence of the formation processes in FEBID the obtained samples are highly disordered.

Electric transport within the metallic grains can be considered diffusive. In general, the grains will have internal defects or defects located at their surface. Trapped charges in the matrix will change the local potential of individual grains. Even if the elastic mean free path inside the grains exceeds the grain diameter, multiple scattering at the grain surface leads to chaotic motion of the electrons which is equivalent to assuming diffusive transport inside the grains due to intra-granular disorder [15]. Nevertheless, the mean spacing δ between the one-electron levels inside the grains is still a well-defined quantity. It is given by $\delta = 1/N_F V$ where $V \sim r^3$ is the grain volume and N_F denotes the density of states at the chemical potential. For grains with a diameter of a few nm, as is typically the case for FEBID structures, δ is of the order of 1 K for metallic grains with density of states of the order of 1 $eV^{-1}nm^{-3}$. Accordingly, quantum size effects due to the discrete energy levels are only relevant at very low temperatures.

The key parameter governing most of the electronic properties of granular metals is the average tunnel conductance G between neighbouring grains. This is most conveniently expressed as a dimensionless quantity $g = G/(2e^2/h)$ by normalization to the quantum conductance. Broadly speaking, metallic behaviour will be observed, if $g \geq 1$, while samples with $g < 1$ show insulating behaviour. The normalized conductivity within a grain is denoted as g_0, not to be confused with the quantum conductance $2e^2/h$, and the notion granular metal implies that $g_0 \gg g$.

Another important parameter is the single-grain Coulomb charging energy $E_C = e^2/2C$ where $C \propto r$ is the capacitance of the grain with radius r. E_C is equal to the change in electrostatic energy of the grain when one electron is added or removed. For insulating samples charge transport is suppressed at low temperatures due to this charging energy. In this respect, the insulating state is closely related to the well-known Coulomb blockade effect of a single grain connected via tunnelling to a metallic reservoir. The average level spacing can become larger than the charging energy for small grains $r < r^*$, where r^* represents the grain radius which separates the regimes for which either the condition $E_C > \delta$ or $E_C < \delta$ holds. In general, the assumption $E_C \gg \delta$ is well justified for nano-granular FEBID structures.

TRANSPORT REGIMES OF NANO-GRANULAR METALS

The transport regimes of granular metals are classified according to the inter-grain coupling strength g. In the strong-coupling limit, $g \gg 1$, a granular array has metallic properties. In the opposite regime $g \ll 1$ the array is insulating. The insulating state is a consequence of the strong Coulomb correlations associated with the single-electron tunnelling. The inter-grain conductance g is best considered as a phenomenological, effective parameter which controls the behaviour of the system. For FEBID structures g cannot be exactly derived from first principles for several reasons, such as the unknown grain size and coupling strength distribution and the ill-defined electronic properties of the matrix.

As temperature is reduced Coulomb correlation and interference effects become important also in the metallic regime. As a consequence, simple Drude-like relations do not hold anymore and the properties of nano-granular metals may differ considerably from those observed in homogeneously disordered metals.

Insulating regime

The insulating regime of granular systems with metallic grains is the regime to which most of the existing work on FEBID structures can be assigned to. If only nearest neighbour single-electron tunnelling is taken into account the conductivity should follow a simple Arrhenius law

$$\sigma(T) \sim \exp[-\Delta/k_B T] \qquad (1)$$

as long as a hard energy gap Δ in the excitation spectrum is present at the chemical potential. Such an activated behaviour is very rarely observed in FEBID structures. Much more frequently the conductivity follows a stretched exponential temperature dependence of the form

$$\sigma(T) = \sigma_0' \exp[-(T_0/T)^{1/2}] \qquad (2)$$

This functional dependence was derived by Efros and Shklovskii for doped semiconductors [16]. Until very recently it remained a puzzle why this same functional form should be obeyed by disordered granular metals in the insulating regime. An early attempt to explain this behaviour based on capacitance disorder due to the grain-size dispersion was discarded because capacitance disorder cannot fully lift the Coulomb blockade of an individual grain, so that a finite gap in the density of states at the chemical potential must remain which would necessarily lead to an Arrhenius behaviour at low temperature [17, 18]. Experimentally, the same stretched exponential was also observed in periodic arrays of quantum dots [19] and periodic granular arrays of gold particles with very small size dispersion [20]. In these systems capacitance disorder was very weak. From this arises the assumption that another type of disorder, unrelated to the grain-size dispersion, is necessary for lifting the Coulomb blockade of a single grain and can lead to a finite density of states at the chemical potential. In recent theoretical work it was suggested that electrostatic disorder, most probably caused by charged defects in the insulating matrix (or substrate), is responsible for lifting the Coulomb blockade [18]. Carrier traps in the insulating matrix at energies below the chemical potential are charged at sufficiently low temperature. They induce a potential of the order $e^2/\varepsilon r$ on the closest granule at a distance r where ε denotes the (static) dielectric constant of the matrix. For two-dimensional granular arrays one can also assume that random potentials are induced by charged defects in the substrate.

At this point it should be remarked that the simple Arrhenius behaviour can be observed in artificial, two-dimensional granular metals prepared by FEBID as detailed in [21, 22]. In these experiments the nano-granular array is formed by taking advantage of the high resolution of the FEBID process in conjunction with using a precursor, $W(CO)_6$, which tends to form near amorphous deposits that can have metal contents of about 40 at%. The diameter of the individual amorphous grains (ca. 20 nm) leads to a well-defined Coulomb blockade which dominates the low-temperature transport properties and causes an Arrhenius behaviour of the conductivity below about 70 K.

Returning to disordered nano-granular FEBID structures it can be stated that in order to fully account for the observed stretched exponential there must be a finite probability for tunnelling to spatially remote states close to the chemical potential. This is analogous to Mott's argument in deriving his variable range hopping (VRH) law for disordered electronic systems in the non-correlated case [23]. Hopping transport over distances exceeding the average distance between adjacent granules in sufficiently dense granular arrays can in principle be realized as tunnelling via virtual electron levels in a sequence of grains, which is also called elastic and inelastic co-tunnelling. This transport mechanism was first considered by Averin and Nazarov as a means to circumvent the Coulomb blockade effect in single quantum dots [24]. In elastic co-tunnelling the charge transfer of a single electron via an intermediate virtual state in an adjacent grain to another state in a grain at a larger distance is at fixed energy. In inelastic co-tunnelling the energies of the initial state and final state differ, so that the electron leaves behind in the granule electron-hole excitations as it tunnels out of the virtual intermediate state. The intermediate states are, as the qualification *virtual* indicates, not classically accessible. The co-tunnelling mechanism was generalized to the case of multiple co-tunnelling through several grains and it was shown that the tunnelling probability falls off exponentially with the distance or the number of grains involved [20, 25, 26]. This is equivalent to the exponentially decaying probability of tunnelling between states near the chemical potential in the theory of Mott, Efros and Shklovskii for doped semiconductors [16, 27] and eventually leads to the expression given in Eq. 2. T_0 is a characteristic temperature which depends on the microscopic characteristics of the granular material in the insulating regime.

The hopping length for transport via virtual electron tunnelling decays as the temperature increases. At some temperature it will be reduced to the average grain size, so that only tunnelling to adjacent grains is possible. The variable range hopping scenario of Mott, Efros and Shklovskii no longer applies. It was suggested that the conductivity should then follow an Arrhenius behaviour [8], as has been observed in ordered granular arrays of gold particles [20] and in artificial nano-dot arrays prepared by FEBID [21, 22].

Metallic regime

As the metal content in FEBID structures increases the situation gets more complicated. In principle one can enter the regime of percolation at a critical metal volume fraction $y = y_c$ in the deposits. Such a percolating path of directly touching metallic grains can short-circuit the remaining part of sample which has activated transport behaviour. Experimentally, this can be studied by a critical exponent analysis of the dependence of the conductivity *vs.* metal volume fraction $\sigma(y)$ at the smallest accessible temperature. In order to estimate whether percolation can play a role for a given volume fraction a suitable micro-structural model has to be applied.

In recent experiments on W-based FEBID structures prepared with the precursor $W(CO)_6$ the behaviour of $\sigma(T)$ shows a qualitative change from activated to non-activated behaviour as the metal content increases [10]. This is in fact indicative of a *insulator-metal transition* as a function of metal content. Apparently, the curvature of $\sigma(T)$ changes sign as the metal-insulator transition is crossed. From these results it can be concluded that percolation of directly touching metallic particles is most likely not the reason for the clear change in the transport mechanism implied by the $\sigma(T)$ behaviour of samples with larger metal content. It is very likely that the microstructure that forms as the metal content is increased is one in which (inelastic) tunnelling prevails to large metal concentrations because the metallic nano-crystals do not touch directly. Rather the tunnelling probability grows with metal content because of grain size growth and a reduction of the intergrain spacing. The crossover to a different transport regime is not simply percolative but is tunnelling, albeit in a more complex form which may need to take into account higher order effects in tunnelling and also correlations. Micro-structural aspects which have to be kept in mind are that the nano-crystals may often have a core-shell structure, with insulating shell, which hinders a direct percolative path to be formed.

A detailed discussion of the present theoretical understanding of the metallic transport regime of granular metals can be found in [8]. As a main result theory predicts in leading order a logarithmic temperature dependence of $\sigma(T)$ independent of the dimensionality of the nano-granular metal. Experimental evidence for this was found transport measurements on annealed Pt-containing FEBID structures [9]. The possible occurrence of higher order corrections of the temperature-dependent conductivity have been discussed in [10]. New and not yet published work of the author's group on the low-temperature conductivity of elec-tron-irradiated nano-granular FEBID structures prepared with the precursor Tri-methyl-methylcyclopentadienyl-platinum MeCpPt(Me)$_3$ give strong indications for the validity of the theoretical predictions.

EXPERIMENTAL EXAMPLE

A particularly nice example of influence of the inter-grain coupling strength on the transport properties of Pt-based nano-granular FEBID structures prepared with the precursor $(CH_3)_3PtC_5H_4CH_3$ (Tri-methyl-methylcyclopentadienyl-platinum) is shown in Figure 2. Here, a series of identical FEBID structures has been irradiated after deposition by 5 keV electrons at 1.6 nA for different periods of time as indicated. As is evident from the plot, for increasing irradiation time the temperature-dependent conductivity shows a cross-over be-haviour from thermally activated towards metallic. After several hours of irradiation this temperature dependence has even changed towards that of a simple metal, i. e. the temperature coefficient of resistance acquires a positive sign which, one could speculate, might be a signature of beginning percolation between the Pt nano-crystallites.

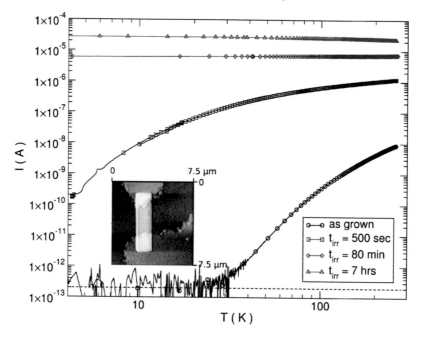

Figure 2. Temperature dependence of the current through Pt-based FEBID structures at a fixed bias voltage of 10 mV. Identical structures show strongly different temperature dependences of their conductance as they have been irradiated in a raster process for different periods of time with electrons at 5 keV and 1.6 nA current. By electron irradiation the FEBID structures can be tuned through a insulator-metal transition. The dashed line indicates the detection limit caused by the finite isolation resistance of the sample wiring. The inset shows one of the as-grown structures between two gold electrodes. The image was acquired by non-contact atomic force microscopy. Height of the deposit is about 120 nm.

Possible reasons for this dramatic change in the conductivity behaviour are irradiation-induced changes in the average Pt grain diameter and/or the properties of the dielectric function of the insulating matrix. At 5 keV beam energy the penetration depth of the electrons into the sensor material amounts to about 120 nm as can deduced from Monte Carlo simulations [28]. For deposits made with the precursor $Pt(PF_3)_4$ volume reduction by loss of phosphor and fluor in conjunction with Pt grain size growth has been reported [29]. For the precursor used in the present case recent transmission electron microscopy investigations gave no indication for a Pt grain size growth under electron irradiation [30]. Our preliminary micro-Raman experiments at 633 nm indicate a change of the dielectric matrix' vibration spectrum from amorphous to nano-crystalline but this needs further elucidation. Presently one is led to speculate that the inter-grain coupling strength growth as a consequence of the electron irradiation driving a insulator-metal transition within a tunnelling-based charge transport regime. If this assumption can be further corroborated by a more detailed analysis of the micro-structural changes brought about by the irradiation process, this kind of irradiation-induced conductivity tuning in nano-granular materials would define a unique and well-controlled handle to studying the correlation-driven metal-insulator transition in disordered systems.

A quite different but innovative aspect is the application of nano-granular materials in the area of strain-sensing by strain-induced changes of the electrical resistance or conductance. Quite generally, the fact that the charge transport is dominated by tunnelling quickly leads to the conclusion that granular metals might be suitable materials for strain-sensing applications, since the tunnel coupling has an intrinsically exponential dependence on the inter-grain distance which is altered under strain; see, e.g., [31] for early work or [32–34] for some recent work on metal-containing diamond-like carbon films.

From the standpoint of a systematic evaluation of the achievable gauge factors κ, i.e., the relative change in resistance normalized to the relative length change in the sensor,

$$\kappa = [\Delta R/R]/[\Delta L/L] \tag{3}$$

little has been done to establish an analysis scheme toward the selection of optimized material parameters which takes the advances in understanding of the charge transport mechanisms in granular metals into account. In actual fact, only recently a theoretical framework has been provided [8, 25] which appears to give full account of the phenomenological similarities in the transport properties of disordered semiconductors and granular metals. In recent work by the author a theoretical methodology for the evaluation of the intrinsic strain dependence of the electrical conductivity of nano-granular metals is introduced. It aims for providing a solid basis for estimating realistically achievable gauge factors for strain sensors based on this material class. Details of this theoretical analysis scheme, which would lead us to far astray at this point, can be found in [11]. In the following section some recent experimental work is presented which highlights some of the favourable properties of nano-granular FEBID structures for strain-sensing applications in the field of micro- and nano-electromechanical systems (MEMS, NEMS).

APPLICATION EXAMPLE – STRAIN-RESISTANCE EFFECT

The field of MEMS and NEMS as enabling technology for sensor device development is rapidly progressing, due to the increasing demand for a continuous down-scaling of sensor functions in different application fields. Different approaches have been followed for nano- and microscale strain/stress measurements ranging from well-established methods, e.g. optical and piezoresistive (see [35, 36] and references therein), to methods still being in their infancy, e.g. carbon nanotubes [37] nanowires [38] and diamond-like carbon films

[39]. Here some very recent results are presented concerning a methodology for strain sensing based on nano-granular metals, using Pt-based deposits as a particularly case study [40]. A specific strength of this methodology is that it does not entail complex fabrication procedures and is readily adaptable to various sensor applications. The high resolution of the FEBID technique allows for easy down-scaling of sensor structures to below 100 nm. The gauge factor for these nano-granular metals depends on the conductivity of the sensor, which can be tuned by electron-beam irradiation leading to a distinct maximum in the sensitivity. By in situ electrical conductivity measurements we are able to tune the sensitivity of the sensor.

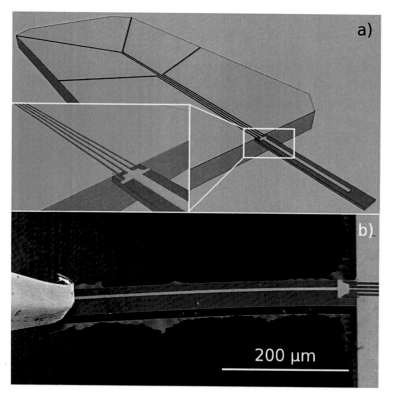

Figure 3. (a) Schematic of the cantilever structure made from silicon with Au contact pads as used in the strain-resistance effect measurements. The length of the cantilever is 500 μm, its height is 10 μm. The zoomed part indicates schematically the position of the FEBID-based strain sensor element. **(b)** SEM image of nano-manipulator tip as it touches the end of the cantilever causing it to bend. At the fixed end of the cantilever three FEBID sensor elements are visible as deposits between the electrodes.

In order to measure the sensitivity of the sensor structures deflection measurements on a cantilever template, as displayed in Figure 3, were performed. The deflection sensitivity for a rectangular cantilever beam which relates the relative resistance change $\Delta R/R$ to the cantilever deflection Δz is given by

$$\Delta R/R \cdot 1/\Delta z = 3\kappa t(1 - L/2)/2l^3 \qquad (4)$$

where κ is the gauge factor of the sensor element, l is the cantilever length, t is the cantilever thickness and L the sensor element length. The sensitivity of the Pt-based sensors was measured by deflecting the cantilever of the sensor chip with a closed loop nano-manipulator while measuring the relative change in resistance as a function of cantilever deflection.

As is shown in Figure 4, the sensor responds with a linear increase in resistance as the cantilever is deflected. The current-voltage characteristic is also linear (see inset) which was verified to hold true up to voltages of more than 5 V, which corresponds to an electric field of more than 2.5 kV/cm.

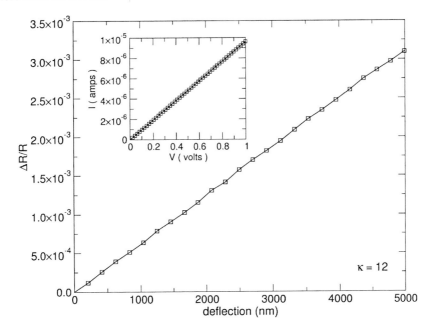

Figure 4. Relative resistance change of a Pt-based cantilever strain-sensor as the cantilever's free end is deflected by Δz. The gauge factor is 12 as indicated. The inset shows the linear current-voltage characteristics of the sensor element. Data taken at room temperature.

For the optimization of a given nano-granular material with regard to the strain-resistance effect the observed dependence of the FEBID structures' conductivity on electron irradiation provides a unique tuning capability. As one example of this effect Figure 5 shows the dependence of the gauge factor on the sensor element's resistivity as the thickness of the sensor element is increased. Apart from the resistance reduction brought about by the increase of the sensor element cross section as the thickness grows, the electron-beam irradiation that accompanies each FEBID process does also contribute to the resistance reduction, as was discussed in the previous section. Apparently, the gauge factor exhibits a pronounced maximum corresponding to a resistance of about 75 kΩ which, assuming a homogenous deposit, amounts to a resistivity of 0.34 Ωcm. A more detailed analysis and attempt to disentangle the pure thickness from the electron irradiation effects can be found in [40].

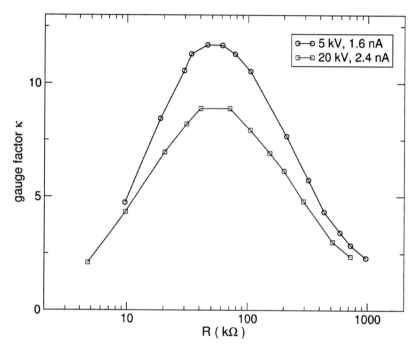

Figure 5. Gauge factor of Pt-based FEBID strain sensors fabricated at different beam conditions as indicated. The measurements were taken in the SEM using a closed-loop nano-manipulator for cantilever deflection and a Wheatstone-bridge setup in conjunction with lock-in technique for detecting the resistance change. The different sensor element resistances are obtained by increasing the thickness of the FEBID deposit. During the growth process the already formed deposit is subject to continuous electron irradiation which causes a much stronger reduction of the resistance than the the pure time-dependent increase of the deposit's height.

CONCLUSION

Focused electron beam deposition has developed in the last decade into a versatile technique for maskless direct beam lithography of functional nano-structures down to the sub-10 nm scale. Present disadvantages for industry-oriented application are the not yet complete control over the metal content in FEBID structures and the need for increasing the deposition yield. It may be concluded that these disadvantages are not inherent to the FEBID process itself but rather reflect the present state of the art in precursor research. With a view to basic science, in particular the field of electronic correlation effects, FEBID does provide a novel methodology for the fabrication of (artificial) nano-granular structures with tunable properties. Recent results on the transport properties of artificial nano-dot arrays fabricated by FEBID suggest that these arrays can provide ideal experimental systems in which to study a variety of interaction-driven quantum phase transitions predicted to occur in Hubbard-like models [41] with typical energy scales in the meV regime, despite of the fact that these models have originally been developed for correlated oxides at typical energy scales of 1 eV. The usability of these correlation effects in strain-sensing applications [11, 40] is yet another aspect that indicates that future work in the field of FEBID processing will likely produce exciting and unexpected results, be it in fundamental or application-driven research.

ACKNOWLEDGMENTS

The author acknowledges financial support by the Beilstein-Institut, Frankfurt am Main, within the research collaboration NanoBiC. Financial support by the NanoNetzwerkHessen (NNH) and by the Bundesministerium für Bildung und Forschung (BMBF) under Grant No. 0312031C is also gratefully acknowledged.

REFERENCES

[1]   Broers, A.N., Molzen, W.W., Cuomo, J. J. and Wittels, N.D. (1976) Electron-beam fabrication of 80 Å metal structures. *Appl. Phys. Lett.* **29**:596. doi: 10.1063/1.89155.

[2]   Matsui, A. and Mori, K. (1984) New Selective Deposition Technology by Electron Beam Induced Surface Reaction. *Jpn. J. Appl. Phys., Part 2* **23**:L706. doi: 10.1143/JJAP.23.L706.

[3]   Matsui, A. and Mori, K. (1986) New selective deposition technology by electron-beam induced surface reaction. *J. Vac. Sci. Technol. B* **4**:299. doi: 10.1116/1.583317.

[4]   Koops, H. W.P., Weiel, R., Kern, D.P. and Baum, T.H. (1988) High-resolution electron-beam induced deposition. *J. Vac. Sci. Technol. B* **6**:477. doi: 10.1116/1.584045.

[5]   Science Citation Index Expanded (SCI-EXPAND).

[6]   Liang, T., Frendberg, E., Lieberman, B. and Stivers, A. (2005) Advanced photolithographic mask repair using electron beams. *J. Vac. Sci. Technol. B* **23**:3101. doi: 10.1116/1.2062428.

[7]   see *e. g.* Okada, S., Mukawa, T., Kobayashi, R., Ishida, M., Ochiai, Y., Kaito, T., Matsui, S. and Fujita, J. (2006) Comparison of Young's Modulus Dependency on Beam Accelerating Voltage between Electron-Beam- and Focused Ion-Beam-Induced Chemical Vapor Deposition Pillars. *Jpn. J. Appl. Phys., Part 1* **45**:5556. doi: 10.1143/JJAP.45.5556.

[8]   Beloborodov, I.S., Lopatin, A.V., Vinokur, V.M. and Efetov, K.B. (2008) Granular electronic systems. *Rev. Mod. Phys.* **79**:469. doi: 10.1103/RevModPhys.79.469.

[9]   Rotkina, L., Oh, S., Eckstein, J.N. and Rotkin, S.V. (2005) Logarithmic behaviour of the conductivity of electron-beam deposited granular Pt/C nanowires. *Phys. Rev. B* **72**:233407. doi: 10.1103/PhysRevB.72.233407.

[10]   Huth, M., Klingenberger, D., Grimm, Ch., Porrati, F. and Sachser, R. (2009) Conductance regimes of W-based granular metals prepared by electron beam induced deposition. *New J. Phys.* **11**:033032. doi: 10.1088/1367-2630/11/3/033032.

[11]   Huth, M. (2010) Granular metals – from electronic correlations to strain-sensing applications. *J. Appl. Phys.* **107**:113709. doi: 10.1063/1.3443437.

[12]   Utke, I., Hoffmann, P. and Melngailis, J. (2008) Gas-assisted focused electron and ion beam processing and fabrication. *J. Vac. Sci. Techn. B* **26**:1197. doi: 10.1116/1.2955728.

[13]   see http://www.pi.physik.uni-frankfurt.de/Sonderforschungsbereich/nanobic/index.html.

[14]   Porrati, F., Sachser, R. and Huth, M. (2009) The transient electrical conductivity of W-based electron-beam-induced deposits during growth, irradiation and exposure to air. *Nanotechnology* **20**:195301. doi: 10.1088/0957−4484/20/19/195301.

[15]   Efetov, K. (1999) Supersymmetry in Disorder and Chaos. Cambridge University Press.

[16]   Efros, A.L. and Shklovskii, B.I. (1975) Coulomb gap and low temperature conductivity of disordered systems. *J. Phys. C* **8**:L 49.

[17]   Pollak, M. and Adkins, C.J. (1992) Conduction in granular metals. *Philosophical Magazine B* **65**:855. doi: 10.1080/13642819208204926.

[18] Zhang, J. and Shklovskii, B.I. (2004) Density of states and conductivity of a granular metal or an array of quantum dots. *Phys. Rev. B* **70**:115317. doi: 10.1103/PhysRevB.70.115317.

[19] Yakimov, A.I., Dvurechenskii, A.V., Nikiforov, A.I. and Bloshkin, A.A. (2003) Phononless hopping conduction in two-dimensional layers of quantum dots. *JETP Lett.* **77**:376. doi: 10.1134/1.1581964.

[20] Tran, T.B., Beloborodov, I.S., Lin, X.M., Bigioni, T. B., Vinokur, V.M. and Jaeger, H.M. (2005) *Phys. Rev. Lett.* **95**:076806. doi: 10.1103/PhysRevLett.95.076806.

[21] Sachser, R., Porrati, F. and Huth, M. (2009) Hard energy gap and current-path switching in ordered two-dimensional nanodot arrays prepared by focused electron-beam-induced deposition. *Phys. Rev. B* **80**:195416. doi: 10.1103/PhysRevB.80.195416.

[22] Porrati, F., Sachser, R., Strauss, M., Andrusenko, I., Gorelik, T., Kolb, U., Bayarjargal, L., Winkler, B. and Huth, M. (2010) Artificial granularity in two-dimensional arrays of nanodots fabricated by focused-electron-beam-induced deposition. *Nanotechnology* **21**:375302. doi: 10.1088/0957-4484/21/37/375302.

[23] Mott, N.F. (1969) Localized states in a pseudogap and near extremities of conduction and valence band. *Philosophical Magazine* **19**:835. doi: 10.1080/14786436908216338.

[24] Averin, D.V. and Nazarov, Yu.V. (1990) Virtual electron diffusion during quantum tunnelling of the electric charge. *Phys. Rev. Lett.* **65**:2446. doi: 10.1103/PhysRevLett.65.2446.

[25] Beloborodov, I.S., Lopatin, A.V. and Vinokur, V.M. (2005) Coulomb effects and hopping transport in granular metals. *Phys. Rev. B* **72**:125121. doi: 10.1103/PhysRevB.72.125121.

[26] Feigelman, M.V. and Ioselevich, A.S. (2005) Variable-range co-tunnelling and conductivity of a granular metal. *JETP Lett.* **81**:277. doi: 10.1134/1.1931015.

[27] Shklovskii, B.I. (1988) Electronic Properties of Doped Semiconductors, Springer, New York.

[28]   Drouin, D., Ré al Couture, A., Joly, D., Tastet, X., Aimez, V. and Gauvin, R. (2007) CASINO V2.42 – A Fast and Easy-to-use Modelling Tool for Scanning Electron Microscopy and Microanalysis Users. *Scanning* **29**:92. doi: 10.1002/sca.20000.

[29]   Botman, A., Hagen, C.W., Li, J., Thiel, B.L., Dunn, K.A., Mulders, J.J.L., Randolph, S. and Toth, M. (2009) Electron postgrowth irradiation of platinum-containing nanostructures grown by electron-beam-induced deposition from $Pt(PF_3)_4$. *J. Vac. Sci. Techn. B* **27**:2759. doi: 10.1116/1.3253551.

[30]   Li, J., Toth, M., Dunn, K.A. and Thiel, B.L. (2010) Interfacial mixing and internal structure of Pt-containing nanocomposites grown by room temperature electron beam induced deposition. *J. Appl. Phys.* **107**:103540. doi: 10.1063/1.3428427.

[31]   Heinrich, A., Gladun, C. and Mönch, I. (1992) Transport properties of composite thin films and application to sensors. *Int. J. Electron.* **73**:883. doi: 10.1080/00207219208925731.

[32]   Takenoa, T., Miki, H. and Takagi, T. (2008) Strain sensitivity in tungsten-containing diamond-like carbon films for strain sensor applications. *Int. J. Appl. Electromagn. Mech.* **28**:211.

[33]   Heckmann, U., Bandorf, R., Gerdes, H., Lübke, M., Schnabel, S. and Bräuer, G. (2009) New materials for sputtered strain gauges. *Procedia Chemistry* **1**:64. doi: 10.1016/j.proche.2009.07.016.

[34]   Koppert, R., Goettel, D., Freitag-Weber, O. and Schultes, G. (2009) Nickel containing diamond like carbon thin films. *Solid State Sci.* **11**:1797. doi: 10.1016/j.solidstatesciences.2009.04.022.

[35]   Zougagh, M. and Rios, A. (2009) Micro-electromechanical sensors in the analytical field. *Analyst* **134**:1274. doi: 10.1039/B901498P.

[36]   Alvarez, M. and Lechuga, M. (2010) Microcantilever-based platforms as biosensing tools. *Analyst* **135**:827. doi: 10.1039/B908503N

[37]   Hierold, C., Jungen, A., Stampfer, C. and Helbling, T. (2007) Nano electromechanical sensors based on carbon nanotubes. *Sens. Actuators A* **136**:51. doi: 10.1016/j.sna.2007.02.007.

[38] Zhou, J., Fei, P., Gu, Y.D., Mai, W. J., Gao, Y.F., Yang, R., Bao, G. and Wang, Z.L. (2008) Piezoelectric-Potential-Controlled Polarity-Reversible Schottky Diodes and Switches of ZnO Wires. *Nano Lett.* **8**:3973. doi: 10.1021/nl802497e.

[39] Takeno, T., Takagi, T., Bozhko, A., Shupegin, M. and Sato, T. (2005) in Metal-containing diamond-like nanocomposite thin film for advanced temperature sensors. *Trans Tech Publications Ltd.* **475-479**:2079 – 2082.

[40] Schwalb, C.H., Grimm, C., Baranowski, M., Sachser, R., Porrati, P., Das, P., Müller, J., Reith, H., Völklein, F., Kaya, A. and Huth, M. (2010) A Tunable Strain Sensor using nanogranular metals. *Sensors* **10**:9847. doi: 10.3390/s101109847.

[41] Stafford, C. A. and Das Sarma, S. (1994) Collective Coulomb Blockade in an Array of Quantum Dots: A Mott-Hubbard Approach. *Phys. Rev. B* **72**:3590. doi: 10.1103/PhysRevLett.72.3590.

Single-Atom Transistors: Atomic-scale Electronic Devices in Experiment and Simulation

Fang-Qing Xie[1,4], Robert Maul[2,3,4],
Wolfgang Wenzel[4,5], Gerd Schön[3,4,5],
Christian Obermair[1,4] and Thomas Schimmel[1,4,5,*]

[1]Institute of Applied Physics, Karlsruhe Institute of Technology (KIT),
Campus South, 76131 Karlsruhe, Germany

[2]Steinbuch Centre of Computing, Karlsruhe Institute of Technology,
Campus North, 76021 Karlsruhe, Germany

[3]Institut für Theoretische Festkörperphysik, Karlsruhe Institute of Technology (KIT),
Campus South, 76131 Karlsruhe, Germany

[4]Center for Functional Nanostructures (CFN), Karlsruhe Institute of Technology
(KIT), 76131 Karlsruhe, Germany

[5]Institute of Nanotechnology, Forschungszentrum Karlsruhe, Karlsruhe Institute of
Technology (KIT), 76021 Karlsruhe, Germany

E-Mail: *thomas.schimmel@physik.uni-karlsruhe.de

Received: 1st September 2010 / Published: 13th June 2011

Abstract

Controlling the electronic conductivity on the quantum level will impact the development of future nanoscale electronic circuits with ultra-low energy consumption. Here we report about the invention of the single-atom transistor, a device which allows one to open and close an electronic circuit by the controlled and reproducible repositioning of one single atom. The atomic switching process is induced by a voltage applied to an independent, third "gate" electrode. In addition to the demonstration of *single*-atom switches, the controlled and reproducible operation of *multi*-atom quantum switches is demonstrated both in experiment and in atomistic calculation. Atomistic modelling of structural and conductance properties elucidates bistable electrode reconstruction as the underlying operation mechanism of the devices. Atomic transistors open intriguing perspectives for the emerging fields of quantum electronics and logics on the atomic scale.

INTRODUCTION

Fascinating physical properties and technological perspectives have motivated investigation of atomic-scale metallic point contacts in recent years [1 – 10]. The quantum nature of the electron is directly observable in a size range where the width of the contacts is comparable to the Fermi wavelength of the electrons, and conductance is quantized in multiples of $2e^2/h$ for ballistic transport through ideal junctions [2].

In metallic point contacts, which have been fabricated by mechanically controlled deformation of thin metallic wires [2 – 4] and electrochemical fabrication techniques [1, 5 – 7] the conductance depends on the chemical valence [2, 3]. Two-terminal conductance-switching devices based on quantum point contacts were developed both with an STM-like setup [8] and with electrochemical methods [9].

In our new approach, a three-terminal, gate-controlled atomic quantum switch was fabricated by electrochemical deposition of silver between two nanoscale gold electrodes (see Fig. 1) [1, 6]. A comparison of the experimental data with theoretical calculations indicates perfect atomic order within the contact area without volume or surface defects [10].

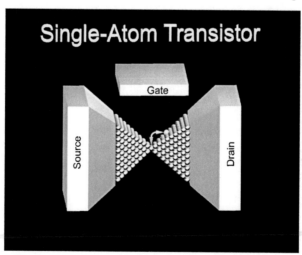

Figure 1. Schematic of the single-atom transistor: The atomic switch is entirely controlled by an independent third gate electrode, allowing to open and close a metallic contact between the source and drain electrodes by the gate-voltage controlled relocation of one single silver atom.

SWITCHING AN ATOM

We control individual atoms in the quantum point contact by a voltage applied to an independent gate electrode, which allows a reproducible switching of the contact between a quantized conducting "on-state" and an insulating "off-state" without any mechanical movement of an electrode (see Fig. 2).

To fabricate the initial atomic-scale contact we deposit silver within a narrow gap between two macroscopic gold electrodes (gap width: typically 50 nm) by applying an electro-chemical potential of $10-40$ mV to a gate electrode [7]. The gold electrodes are covered with an insulating polymer coating except for the immediate contact area, and serve as electrochemical working electrodes. They correspond to the "source" and "drain" electrodes of the atomic-scale transistor. Two silver wires serve as counter and quasi-reference electrodes.

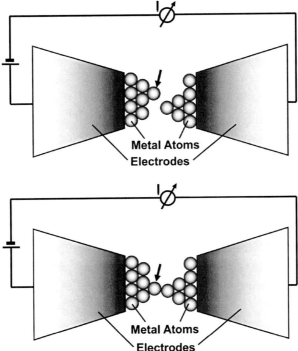

Figure 2. Schematic of the switching process: A metal atom (see arrow) is switched between a quantized conducting "on-state" (lower graph) and an insulating "off-state" (upper graph).

The potentials of the working electrodes with respect to the quasi-reference and counter electrodes are set by a computer-controlled bipotentiostat (see Fig. 3). All experiments are performed at room temperature, the electrolyte being kept in ambient air. For conductance measurements, an additional voltage in the millivolt range is applied between the two gold electrodes. To fabricate the atomic transistor, silver is deposited on each of the two working electrodes, until finally two silver crystals meet, forming an atomic scale contact which is bridging the gap.

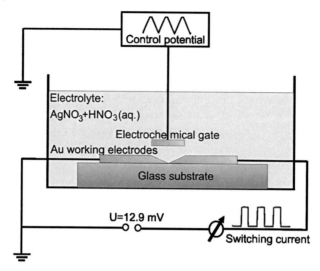

Figure 3. The experimental setup. Within a narrow gap between two gold electrodes on a glass substrate, a silver point contact is deposited electrochemically. By a procedure involving repeated computer-controlled electrochemical cycling, a bistable atomic-scale quantum conductance switch is fabricated.

While silver islands are deposited in the junction we monitor the conductance between the two electrodes. As soon as the conductance exceeds a preset "target" value, the deposition is stopped and the voltage is reversed to dissolve the junction again. After the conductance drops below a preset value, the deposition/dissolution cycle is repeated automatically by the computer-controlled setup. During the first cycles, the conductance at the end of the deposition step varies strongly from cycle to cycle [11]. After repeated cycling, an abrupt change is observed from this irregular variation to a bistable switching between zero and a finite, quantized conductance value at an integer multiple of G_0 ($= 2e^2/h$).

Controlling the junction at the single-atom level

Figure 4 shows a sequence of reproducible switching events between an insulation "off-state" and a quantized conducting "on-state" (at 1 G_0), where the quantum conductance (red curves) of the switch is controlled by the gate potential (blue curves), as commonly observed in transistors. As calculations have shown [10], for atomic-scale silver contacts a quantized conducting "on-state" of 1 G_0 corresponds to a *single-atom* contact.

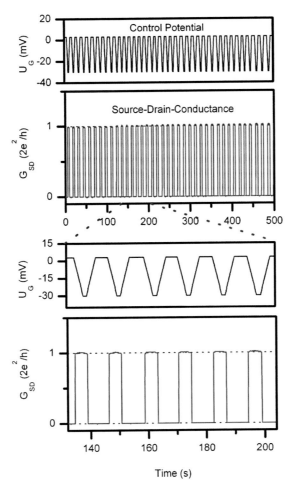

Figure 4. Switching an electrical current by gate-controlled atomic movement: Experimental data of reproducible electrical switching with a single silver atom point contact between an "on-state" at 1 G_0 (1 $G_0 = 2e^2/h$) and a non-conducting "off-state". The source-drain conductance (G_{SD}) of the atomic switch (red curves) is directly controlled by the gate potential (U_G) (blue curves).

When we set the gate potential to an intermediate "hold" level between the "on" and the "off" potentials, the currently existing state of the atomic switch remains stable, and no further switching takes place. This is demonstrated in Figure 5 both for the "on-state" of the switch (left arrow) and for the "off-state" of the switch (right arrow). Thus, the switch can be reproducibly operated by the use of three values of the gate potential for "switching on", "switching off" and "hold". These results give clear evidence of a hysteresis when switching between the two quantized states of the switch. It can be explained by an energy barrier which has to be overcome when performing the structural changes within the contact when switching from the conducting to the non-conducting state of the switch and *vice versa*.

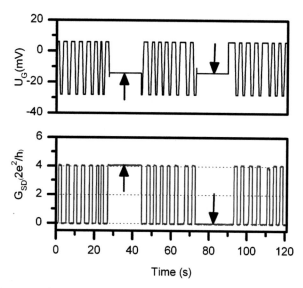

Figure 5. Demonstration of quantum conductance switching between a non-conducting "off-state" and a preselected quantized "on-state" at 4 G_0. A conductance level can be kept stable, if U_G is kept at a "hold" level (see arrows) (cf. [1]).

The results indicate that switching occurs by a reversible rearrangement of the contacting group of atoms between two different stable configurations with a potential barrier between them. For silver the observed quantum conductance levels appear to coincide with *integer* multiples of the conductance quantum [1, 10].

The observed integer conductance levels of the switch are determined by the available bistable junction conformations, similar to the observation of preferential atomic configurations in metallic clusters corresponding to "magic numbers" [12]. By snapping into 'magic' bistable conformations, such energetically preferred junction configurations are mechanically and thermally stable at room temperature, and they are reproducibly retained even during long sequences of switching cycles.

Multilevel Switching

Reproducible switching in the above cases was always performed by opening and closing a quantum point contact, i.e. by switching between a quantized conducting state and a non-conducting state. However, it was not clear if this kind of gate-electrode controlled switching is also possible between two different conducting states of one and the same contact. Such kind of switching would involve two different stable contact configurations on the atomic scale, between which reversible switching would occur even without ever breaking the contact.

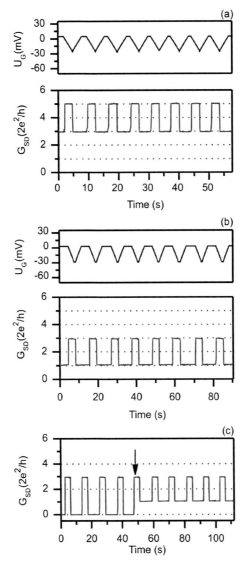

Figure 6. (a) and **(b)**: Experimental demonstration of the operation of an *interlevel* transistor based on an atomic-scale silver quantum point contact. A controlled change of the electrochemical gate potential U_G leads to controlled switching of the conductance of the quantum point contact G_{SD} between *two different* quantized conducting states each of which is different from zero. The upper diagram gives the control potential (blue curve) applied to a "gate" electrode while the lower shows the corresponding conductance switching of the point contact (red curve). The two states between which the switching occurs in **(a)** exhibit conductance levels of 3 G_0 $(G_0 = 2e^2/h)$ and 5 G_0, respectively. **(b)** Example of another *interlevel* quantum transistor switching between the conductance levels of 1 G_0 and 3 G_0. **(c)** Experimental demonstration of a multi-level atomic-scale transistor switching between an "off-state" and two different "on-states" (cf. [13]).

Such multi-level logics and storage devices on the atomic scale would be of great interest as they allow a more efficient data storage and processing with a smaller number of logical gates. By developing a modified procedure of fabrication, a multi-level atomic quantum transistor was obtained, allowing the gate-controlled switching between *different* conducting states.

Instead of setting the lower threshold where the dissolution process is stopped by the computer, to a value near 0 G_0, the lower threshold was set at a value above the desired quantized conductance of the lower of the two "on-state" levels [13].

Figure 6 demonstrates the operation of such a two-level transistor: A controlled change of the gate potential U_G leads to a controlled switching of the conductance of the quantum point contact between two *different* quantized conducting states (for details see caption of Fig. 6). Sharp transitions are observed between the two levels. No intermediate steps or staircase-like structures in conductance are observed in the diagram. The transitions are instantaneous within the time resolution of the diagram of Figure 6 (50 ms).

COMPUTER SIMULATIONS OF THE SWITCHING PROCESS

Reproducible switching between quantum conductance levels over many cycles cannot be explained by conventional atom-by-atom deposition but requires a collective switching mechanism. Our calculations have shown that only well-ordered junction geometries result in integer multiples of the conductance quantum [10]. Neither partial dissolution of the junction nor its controlled rupture yields the necessary atomic-scale memory effect. A more detailed model of the structural [15, 16] and conductance [3, 17] properties of such junctions is therefore required.

We begin with the unbiased deposition of silver ions (Fig. 7a), which evolve under the influence of the electrostatic potential of the electrodes [18]. Starting from two distant parallel Ag(111) layers, we evolve each ion in the long-range electrostatic potential generated from the present electrode conformation and a short-range Gupta potential for silver [19]. In the simulations, we deposit one atom at a time using a kinetic Monte Carlo method [18] starting at a random position inside the junction. We deposit up to 800 atoms in the junction until a predefined number of non-overlapping pathways connect the left and right electrode. As a non-overlapping pathway, we define a unique set of touching atoms that extend from one electrode to the other, which permits us to identify the minima cross-section of the junction.

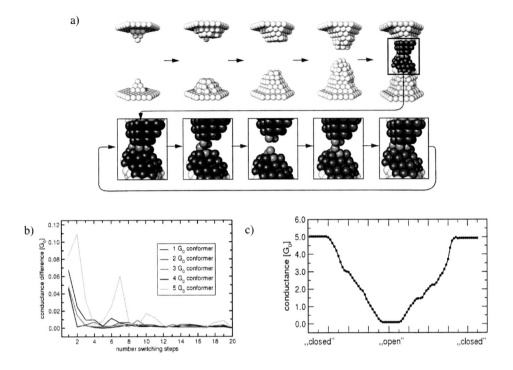

Figure 7. Simulation of atomic point contact growth and switching process. **(a)** Snapshots of the deposition simulation. Upper row: The growth process starts with two disconnected Ag(111) layers and stops, when a non-overlapping pathway with a predefined number of silver atoms connects the electrodes. Lower row: Simulation of the switching process reveals a bistable tip reconstruction process as the mechanism underlying the reproducible switching of the conductance. During the simulation, we kept the silver atoms marked in gray at their positions at the end of the deposition and permitted the central cluster of atoms to evolve (blue and red atoms) under the influence of the electrochemically induced pressure. The central silver atoms (red) define the minimal cross-section. These atoms return with sub-Angstrom precision to their original positions at the end of the switching cycle. **(b)** Difference in the computed conductance between subsequent "on" conformations as a function of the switching cycle for selected junctions of 1 G_0, 2 G_0, 3 G_0, 4 G_0, and 5 G_0, respectively. Junctions switch reproducibly for over 20 cycles between increasingly stable "on" and "off" conformations (training effect). **(c)** Variation of the computed conductance of a 5 G_0 switch during one "open-close" process. In agreement with the experimental observations, we find asymmetric plateaus in the conductance curve, if the switch is opened or closed. This can be traced back to the existence of several low-energy pathways connecting the open and closed state (cf. [1]).

Next, we simulate the switching process: The change in the electrochemical potential induces a change in the interface tension of the liquid-metal interface, making possible a deformation of the junction geometry parallel to the junction axis. It is well-known that changes in the electrochemical potential modulate the interfacial tension of the whole electrode [20−22] which results in a mechanical strain on the junction. We simulate the opening/closing cycle of a junction by evolving the atoms of a "central" cluster under the influence of the electrochemical pressure. We assume that only the atoms in this region move in the switching process, while most of the bulk material remains unchanged. The central cluster comprises the atoms of the minimal cross-section connecting the two electrodes and all atoms within a radius of 9.0 Å around this central bottleneck. While the electrodes gradually move apart/closer together, all atoms of the central cluster relax in simulated annealing simulations generating a quasi-adiabatic path between the open and the closed conformation.

Not surprisingly, the junction rips apart at some finite displacement from the equilibrium, an effect also seen in break-junction experiments. For most junctions, this process is accompanied by a surface reorganization on at least one, but often both, tips of the electrode(s). When we reverse the process, some junctions snap into the original atomistic conformation with subatomic precision. Figure 8 shows the simulated contact geometries (b) and the related minimum cross sections (c) for 1...5 G_0 together with the corresponding experimentally determined switching curves (a). At the end of the switching simulation, we compare the final and the starting geometry. If after the first switching cycle the junction has returned to the same geometry, we consider the junction "switchable" and perform further switching cycle simulations to test stability. Otherwise, we discard the junction completely and start from scratch.

We have performed 15280 full deposition simulations generating $N_{conf} = 17, 8, 3, 17$, and 6 junctions with $p = 1,..., 5$ conductance quanta, respectively. Most deposition simulations fail to generate a switchable junction because the "acceptance criterion" for "switchability" was very strict. We note that the same holds true for most control simulations starting from the "perfect" conformations of [10], indicating that simple rupture of even nearly ideal junctions cannot be the basis of the switching mechanism.

We then compute the zero-bias conductance [23 – 25] of the entire junction using a material-specific, single particle Hamiltonian and realistic electrode Green's functions [26]. We use a recursive Green's function method [27, 28] which maps the problem of computing the full device Green's function to the calculation of "principal layer" Green's functions, which drastically reduces the computational effort but maintains the accuracy. The electronic structure is described using an extended Hückel model including s-, p-, and d-orbitals for each silver atom (7200 orbitals per junction [29] in the standard minimal basis set of non-orthogonal Slater type orbitals. We take the influence of the leads into account, by assuming a semi-infinite fcc lattice for the left and the right reservoir. We compute the material-specific surface Green's functions by applying a decimation technique that exploits the translational symmetry of the semi-infinite contacts [30]. We find that the retained junction conformations (typically comprising 500 – 800 atoms) have a preselected integer multiple (n) of G_0 in close agreement with the experiment. The direct comparison of our atomistic, quantum conductance calculations, using the unaltered conformations from the deposition/

switching simulations, with the experimental conductance measurements offers a strong validation of the geometries generated in our deposition protocol. The observed agreement between computed and measured conductance is impressive, because the conductance of metallic wires is well-known to be strongly dependent on the geometry.

We have repeated this process up to 20 times for each junction (Fig. 7b) and observe a "training effect", in which the junction geometries become increasingly stable, alternating between two bistable conformations. When re-computing the zero-bias conductance at the end of switching cycle, we find the same value as for the original junction (to within ~0.05 G_0). Because these observations result from completely unbiased simulations of junction deposition and switching, they explain the observed reversible switching on the basis of the generation of bistable contact geometries during the deposition cycle. If we consider the tip-atoms at each side of the electrodes in the open junction, the equilibrium geometry of both clusters depends on their environment. In the open junction, this environment is defined by the remaining electrode atoms on one side, while in the closed junction, the tip-cluster of the other electrode is also present. Our simulations demonstrate the existence of two stable geometries for each cluster in both environmental conditions, respectively. Reversible switching over many cycles is thus explained by reversible tip reorganization under the influence of the gate potential, similar to induced surface reorganization [31 – 33]. While the overall structure differs between junctions with the same conductance quantum from one realization to the next (see Fig. 8b for representative examples), the minimal cross-section that determines the conductance is largely conserved (Fig. 8c). When comparing the opening process and the closing process of the junction, we observe asymmetric conductance curves (see Fig. 7c), in agreement with experiment, resulting from irreversible low-energy pathways between the open and the closed conformations.

These data rationalize the bistable reconfiguration of the electrode tips as the underlying mechanism of the formation of nanoscale junctions with predefined levels of quantum conductance. These levels are determined by the available bistable junction conformations, similar to magic numbers for metal clusters [33] that are most likely material-specific. For silver, the observed quantum conductance levels appear to coincide with integer multiples of the conductance quantum. When we form a junction by halting the deposition process at a non-integer multiple of G_0 (both experimentally and in simulation), subsequent switching cycles either converge to an integer conductance at a nearby level or destroy the junction. By snapping into "magic" bistable conformations, junctions are mechanically and thermally stable at room temperature for long sequences of switching cycles. This process is assisted by the electrochemical environment but not intrinsically electrochemical: The reproducible switching of large junctions by coordinated dissolution/regrowth of the junction is very unlikely.

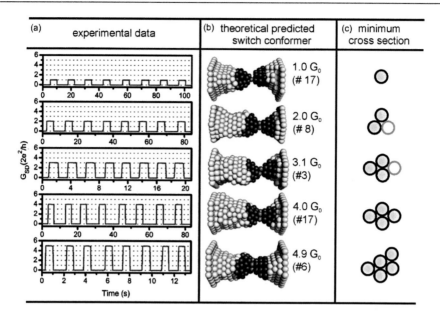

Figure 8. Relation between the structures of atomic point contacts and their conductance. **(a)** Quantum conductance switching between a non-conducting "off-state" and a preselected quantized "on-state" at 1 G_0, 2 G_0, 3 G_0, 4 G_0, and 5 G_0, respectively (note individual time axis). **(b)** Representative conformations of simulated junctions, computed zero-bias conduction, and number of junctions with the specified conductance. **(c)** Representative minimal cross-sections for each conductance level. The minimal cross-sections are characteristic for each group of the switch conformers and determine their quantized conductance (cf. [1]).

CONCLUSIONS AND PERSPECTIVES

The development of the single-atom transistor represents a first demonstration of the functionality of a transistor on the atomic scale. This is of great interest, as there were many previous demonstrations of passive devices such as atomic-scale and molecular resistors. However, there was a lack of an actively switching device such as a transistor on the atomic scale. The atomic transistor as an actively controllable device, which reproducibly operates at room temperature, is filling this gap.

The switching process is explained by a reproducible rearrangement of individual atoms between two well-defined geometries. Computer simulations are in excellent agreement with the experimental results and show the self-stabilization of the switching process.

Atomic transistors represent a new class of devices which show remarkable properties:

- They allow the switching of an electrical current by the *geometrical relocation of individual atoms* rather than by locally changing electronic properties as done in conventional transistors.

- They represent *quantum switches*, the levels between which the switching occurs being given by fundamental laws of quantum mechanics.

- They are a first demonstration of an *all-metal transistor* without the use of any semiconductor, the lack of a band gap allowing operation at very low voltages.

Such devices provide a number of advantages: They possess extremely nonlinear current-voltage characteristics, desirable in many applications, and they can be manufactured using conventional, abundant, inexpensive and non-toxic materials. At the same time, the devices open perspectives for electronic switching at ultrafast frequencies: although the switching time in our current investigations is limited by the response time of the electrodes ($3-5$ microseconds), the intrinsic operation time is expected to be limited by the atomic-scale rearrangement within the junction (picoseconds), opening perspectives for ultrahigh frequency operation. Because the switching process is achieved with very small gate potentials in the millivolt range, the power consumption of such devices is by orders of magnitude lower than that of conventional semiconductor-based electronics.

Although the development of the single-atom transistor just marks the beginning of actively controlled electronics on the atomic scale, it opens fascinating perspectives for quantum electronics and logics based on individual atoms. The development of a first, simple integrated circuit [1, 14] and a multilevel quantum transistor [13] are first encouraging steps in this direction.

ACKNOWLEDGMENT

This work was supported by the Deutsche Forschungsgemeinschaft within the DFG Centre for Functional Nanostructures and by the Baden-Württemberg Foundation within the Network of Excellence on Functional Nanostructures, Baden-Württemberg. Part of the experimental section were reproduced with permission of *Europhysics News* (2010) [34], Figure 6 was reproduced with permission from *Advanced Materials* (Copyright 2010, Wiley-VCH) [13], Figures 7 and 8 with permission from *Nano Lett.* (ACS, Copyright 2010) [1].

REFERENCES

[1]   Xie, F.-Q., Maul, R, Augenstein, A, Obermair, Ch., Starikov, E.B., Schön, G., Schimmel, Th., Wenzel, W. (2008) *Nano Lett.* **8**:4493.
doi: 10.1021/nl802438c.

[2]   Agraït, N., Yeyati, A.L., van Ruitenbeek, J.M. (2003) *Phys. Rep.* **377**:81.
doi: 10.1016/S0370-1573(02)00633-6.

[3]   Scheer, E., Agraýt, N., Cuevas, N.J., Levy Yeyati, A., Ludoph, B., Martin-Rodero, A., Rubio Bollinger, G., van Ruitenbeek, J.M., Urbina, C. (1998) *Nature* **394**:154.
doi: 10.1038/28112.

[4]   Mares, A.I., van Ruitenbeek, J.M. (2005) *Phys. Rev. B* **72**:205402.
doi: 10.1103/PhysRevB.72.205402.

[5]   Li, C. Z., Bogozi, A., Huang, W., Tao, N.J. (1999) *Nanotechnology* **10**:221.
doi: 10.1088/0957-4484/10/2/320.

[6]   Xie, F.-Q., Nittler, L., Obermair, Ch., Schimmel, Th. (2004) *Phys. Rev. Lett.* **93**:128303.
doi: 10.1103/PhysRevLett.93.128303.

[7]   Xie, F.-Q., Obermair, Ch., Schimmel, Th. (2004) *Solid State Communications* **132**:437.
doi: 10.1016/j.ssc.2004.08.024.

[8]   Smith, D.P.E. (1995) *Science* **269**:371.
doi:10.1126/science.269.5222.371.

[9]   Terabe, K., Hasegawa, T., Nakayama, T., Aono, M. (2005) *Nature* **433**:47.
doi: 10.1038/nature03190.

[10] Xie, F.-Q., Maul, R., Brendelberger, S., Obermair, Ch., Starikow, E.B., Wenzel, W., Schön, G., Schimmel, Th. (2008) *Appl. Phys. Lett.* **93**:043103.
doi: 10.1063/1.2955521.

[11] Xie, F.-Q., Obermair, Ch., Schimmel, Th. (2006) *In*: R. Gross *et al.* (eds.), *Nanoscale Devices - Fundamentals and Applications.* Springer, p153.

[12] Huda M.N., Ray, A.K., (2003) *Phys. Rev. A* **67**:013201.
doi: 10.1103/PhysRevA.67.013201.

[13] Xie, F.-Q., Maul, R., Obermair, Ch., Schön, G., Wenzel, W., Schimmel, Th. (2010) *Advanced Materials* **22**:2033.
doi: 10.1002/adma.200902953.

[14] Schimmel, Th., Xie, F.-Q., Obermair, Ch. Patent pending, US 2009195300.

[15]   Pauly, F., Dreher, M., Viljas, J.K., Häfner, M., Cuevas, J.C., Nielaba, P. (2006) *Phys. Rev. B* **74**:235106.
       doi: 10.1103/PhysRevB.74.235106.

[16]   Yanson, I.K., Shklyarevskii, O.I., Csonka, S., van Kempen, H., Speller, S., Yanson, A.I., van Ruitenbeek, J.M. (2005) *Phys. Rev. Lett.* **95**:256806.
       doi: 10.1103/PhysRevLett.95.256806.

[17]   Yanson, A.I., Yanson, I.K., van Ruitenbeek, J.M. (2001) *Phys. Rev. Lett.* **87**:216805.
       doi: 10.1103/PhysRevLett.87.216805.

[18]   Kwiatkowski, J.J., Nelson, J., Li, H., Bredas, J.L., Wenzel, W., Lennartz, C. (2008) *Phys. Chem. Chem. Phys.* **10**:1852.
       doi: 10.1039/b719592c.

[19]   Shao, X., Liu, X., Cai, W.J., (2005) *J. Chem. Theory Comput.* **1**:762.
       doi: 10.1021/ct049865j.

[20]   Weissmuller, J., Viswanath, R.N., Kramer, D., Zimmer, P., Wurschum, R., Gleiter, H. (2003) *Science* **300**:312.
       doi: 10.1126/science.1081024.

[21]   Weigend, F., Evers, F., Weissmüller, J. (2006) *Small* **2**:1497.
       doi: 10.1002/smll.200600232.

[22]   Inglesfield, J.E. (1985) *Prog. Surf. Sci.* **20**:105.
       doi: 10.1016/0079-6816(85)90007-3.

[23]   Xue, Y., Datta, S., Ratner, M. (2002) *Chem. Phys.* **281**:151.
       doi: 10.1016/S0301-0104(02)00446-9.

[24]   Heurich, J., Cuevas, J.C., Wenzel, W., Schön, G. (2002) *Phys. Rev. Lett.* **88**:256803.
       doi: 10.1103/PhysRevLett.88.256803.

[25]   van Zalinge, H., Bates, A., Schiffrin, D.J., Starikov, E.B., Wenzel, W., Nichols, R.J. (2006) *Angew. Chem.* **45**:5499.
       doi: 10.1002/anie.200601263.

[26]   Jacob, T. (2007) *Electrochim. Acta* **52**:2229.
       doi: 10.1016/j.electacta.2006.03.114.

[27]   Vergés, J.A. (1999) *Comput. Phys. Commun.* **118**:71.
       doi: 10.1016/S0010-4655(99)00206-4.

[28]   Maul, R., Wenzel, W. (2009) *Phys. Rev. B* **80**:045424.
       doi: 10.1103/PhysRevB.80.045424.

[29]   Starikov, E.B., Tanaka, S., Kurita, N., Sengoku, Y., Natsume, T., Wenzel, W. (2005) *Eur. Phys. J. E* **18**:437. doi: 10.1140/epje/e2005-00047-4.

[30]   Damle, P., Ghosh, A.W., Datta, S. (2002) *Chem. Phys.* **281**:171. doi: 10.1016/S0301-0104(02)00496-2.

[31]   Ohiso, A., Sugimoto, Y., Abe, M., Morita, S. (2007) *Jpn. J. Appl. Phys.* **46**:5582. doi: 10.1143/JJAP.46.5582.

[32]   Ternes, M., Lutz, C.P., Hirjibehedin, C.F., Giessibl, F.J., Heinrich, A.J. (2008) *Science* **319**:1066. doi: 10.1126/science.1150288.

[33]   Huda, M.N., Ray, A.K. (2003) *Phys. Rev. A* **67**:013201. doi: 10.1103/PhysRevA.67.013201.

[34]   Obermair, Ch., Xie, F.-Q., Schimmel, Th. (2010) *Europhysics News* **41**:25. doi: 10.1051/epn/2010403.

FUNCTIONAL NANOSCIENCE: PRESENT AND FUTURE

DAVID WINKLER

CSIRO Materials Science and Engineering,
Bag 10, Clayton South MDC 3169, Australia.

E-MAIL: dave.winkler@csiro.au

Received: 1st September 2010 / Published: 13th June 2011

FIFTY YEARS AFTER FEYNMAN

There have been many influences and drivers for the development of technologies that allow functional components to be constructed at smaller and smaller scale. The semiconductor revolution in the second half of the 20th century was driven by cost, speed, novel function, and power consumption. Semiconductor science and its child, large-scale integration of electronic circuitry, have been responsible for an unprecedented paradigm change in almost every aspect of human life. The change is arguably even more profound than that which resulted from the industrial revolution. As we shall see later in this paper, although the fundamental limits of Moore's Law have not yet been reached, this and the increasing energy consumption of these paradigm-breaking technologies will necessitate another paradigm shift in the near future.

In terms of the influence of individuals, the development of what we now call functional nanoscience clearly owes much to several outstanding scientists, all of whom were awarded the Nobel Prize for their work. Shockley, Bardeen and Brattain's discovery of the transistor, Kilby's invention of the integrated circuit, Krug's development of electron microscopy, Watson, Crick, and Wilkins' discovery of the structure, self-assembly of, and information processing in DNA, Prigogine's work on self-organization in dissipative structures, Cram, Lehn and Pedersen's development of self-assembled molecular structures, Smalley, Kroto and Curl's seminal discovery of buckminsterfullerenes, and Boyer and Walker's discovery of that archetypal molecular machine, ATP synthase, have all been major drivers for, and facilitators of, contemporary nanoscience.

However, it is another Nobel Laureate, Richard Feynman, who widely credited with providing essential stimulus to the development of functional nanoscience. His famous lecture, *There's Plenty of Room at the Bottom, An Invitation to Enter a New Field of Physics* was delivered at the annual meeting of the American Physical Society at the California Institute of Technology (Caltech) in December 1959. The lecture was first published in the February 1960 issue of Caltech's *Engineering and Science* [1]. It challenged scientists to think about constructing devices at the nanoscale, a possibility now brought to fruition in the early 21st century. The current paper is a summary of a series of truly inspirational and engrossing lectures, describing the state of the art of functional nanoscience, presented at the Beilstein Bozen Symposium on Functional Nanoscience. It attempts to draw together common themes in quite diverse areas of functional nanoscience, from self-assembled molecular devices and nanoscale cantilevers, through bio-inspired materials and molecular machines, to quantum electronics and lithography. Although Niels Bohr warned of making predictions, especially about the future, we also attempt to anticipate where functional nanoscience is leading in the short, medium, and long term.

SYSTEMS CHEMISTRY AND FUNCTIONAL NANOSCIENCE

As the theme of the preceding Beilstein Symposium was systems chemistry [2], it is instructive to show how this new field of science overlaps with and informs functional nanoscience. Although systems chemistry is currently poorly defined, one definition is that it is the application of complex systems science to chemistry and molecular science. Complex systems science is playing an increasingly important role as an alternative, complementary approach to reductionism. Most physical and biological systems are intrinsically complex, consisting of a myriad of low-level components interacting to generate overt system (emergent) properties. Complex systems science approaches to chemistry have been reviewed recently [2]. Key elements of complexity theory are self-organization, criticality, emergent properties, and the description of interacting systems as networks. Networks describe the interactions between components of very complicated systems. Although interactions between system components may be relatively simple, they are nonetheless capable of generating extremely complex and sophisticated behaviour of the system as a whole. Network architecture influences the robustness and information flow of networks. In many robust networks commonly favoured by biological systems as diverse a social and ecological networks, gene and protein interaction networks, highly connected network nodes (hubs) play an important role in controlling network behaviour. The properties of the hubs (connectivity, etc.) and the architecture of the network in which they are embedded, contribute to network robustness and vulnerability. Hubs occur in social networks, such as citation and collaboration networks, where key researchers are highly cited and have a large number of collaborations, substantially influencing social networks of researchers in specific fields.

An example of an influential 'hub' scientist in chemistry is George Whitesides whose work has been a significant influence for some of the work presented at the Symposium. Analysis of Whitesides collaboration and citation network shows the extent of his influence, amplifying his ability to be an "ideas catalyst" in several major chemistry-related fields. In functional nanoscience and in the context of the Beilstein Symposium, influential hub scientists are exemplified by Nadrian Seeman (New York University, NY, U.S.A.), whose DNA origami technique is used in over 60 laboratories, and by Christoph Gerber's (University of Basel, Switzerland) seminal and hugely influential developments of the atomic force microscope (AFM), and scanning tunnelling microscope (STM)).

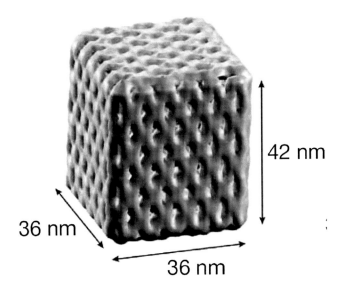

Figure 1. Self-assembled DNA cube.

Like scientists, molecules can also be influential hubs in research fields. In functional nanoscience molecular hubs include buckminsterfullerenes, nanotubes, and DNA. We shall see later in this chapter when discussing the work of Keith Firman (University of Portsmouth, UK), Nadrian Seeman, Jørgen Kjems (Åarhus University, Denmark), and William Shih (Harvard Medical School, Boston, MA, U.S.A.) the pivotal role that DNA plays in design and self-assembly of precisely assembled molecular objects.

Complex systems science concepts like self-organization, self-assembly, emergent properties, criticality, and network properties will play increasingly important roles in functional nanoscience in the future.

THE PRESENT

Chemists would argue that we have been doing 'nanoscience' for a long time. Although it is perhaps the ultimate reductionist science, chemistry has developed a remarkable suit of capabilities over the past century. Current chemical synthesis methods allow exceedingly complex molecules to be assembled via precise control of atom placement, as the total synthesis of ciguatoxins (Figure 2) by Hirama illustrates [3]. Essentially any atom that obeys the laws of chemical reactivity and valence can be placed at will in any position to obtain complex synthetic molecules, which are often designed to have a specific purpose. The main limitations are time and cost.

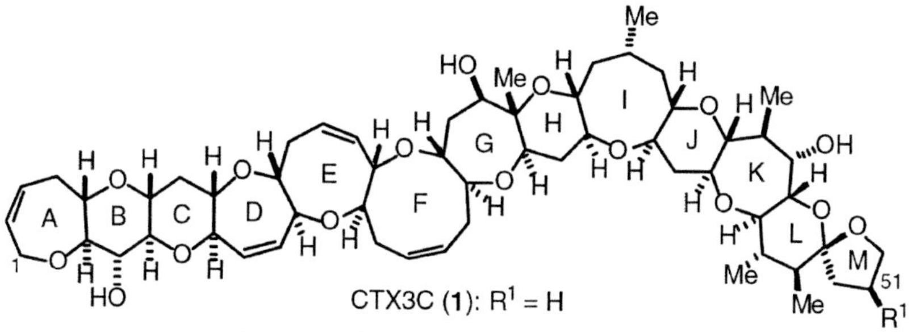

CTX3C (1): R¹ = H

51-hydroxyCTX3C (2): R¹ = OH

Figure 2. Structures of ciguatoxins.

Nanoscience draws heavily on organic chemistry, colloid and surface science, inorganic chemistry, molecular biochemistry, and macromolecular chemistry in particular. These fields have provided a source of knowledge about biological nanomachines, and the means for constructing, visualizing and characterizing very small objects. Functional nanoscience, *the ability to create, move and image nanoscale objects*, is now moving from the era of 'passive' nanoparticles, towards 'active' nanomachines with multiple functions and capabilities, increasing autonomy and complexity.

Contemporary functional nanoscience can be essentially subdivided and several different ways. Top down and bottom up nanoscience generally describe nanodevices that are engineered either by micromachining of a larger substrate material, or that self-assemble spontaneously via precisely designed molecular interactions as in biological systems for example.

Like all new, exciting scientific fields, nanoscience stimulates imaginative and sometimes overly optimistic or fanciful images of its possibilities. This is captured in the graph in Figure 3 that shows the history of scientific discoveries. The period of exaggerated expectations is followed, when these high expectations go unmet, with a period of disappointment. A return to the fundamentals of the technology allows science to link with applications and real commercial investment begins [4].

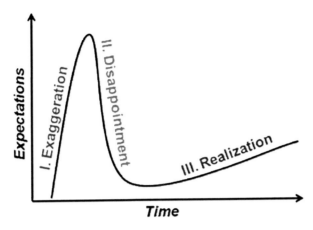

Figure 3. The expectations of a new technology as a function of time [4].

In this context, it is interesting to revisit the initial expectations of functional nanoscience as summarized in the recent review by Whitesides and Lipomi [4].

- revolutionary electronics (including devices constructed from single organic molecules)
- ultra-dense microprocessors and memories, and quantum computers;
- futuristic speculations concerning nano robots and nanoscale machines;
- revolutionary materials with extraordinary applications (*e.g.*, buckytubes and the space elevator, or quantum dots and cancer-targeted drugs);
- applications in biomedicine relying on particles small on the scale of the cell.

We can now contrast these with the nanoscience applications that have emerged [4].

- information technology and nanoelectronics. Developments in conventional microfabrication, like short-wavelength light sources and immersion optics, have moved microfabrication into the deep nanoscale region (~45 nm). Extreme ultraviolet lithography, double patterning lithography, and multiple maskless electron-beam writing will push the limit lower.
- an enormous range of opportunities for nanoscience in energy production, These range from heterogeneous catalysis to improved membranes for separation of water and gases

- soft nanotechnology in biology. The role of nanoscience in fundamental biology is still to be established, but is discussed at length in this paper.

- the physical chemistry of systems containing small numbers of molecules or particles, and having most of these particles on or close to a surface, is just beginning to be explored.

- quantum effects that emerge in small systems are very much unexplored.

Where is functional nanoscience now? Passive nanodevices have essentially formed the bulk of past and contemporary nanoscience research. They are generally particles that have controllable properties but are still essentially 'simple' with no inbuilt intelligence, energy source, or means of propulsion. Passive nanoscience is approaching maturity. There were > 1000 nano products on market in U.S. at the end of 2009 [5].

The other, more challenging class of nanodevices that form the theme of this workshop are the active or functional nanodevices. Compared with the passive nanodevices, the field is embryonic but is developing rapidly and has great promise.

ISSUES/CONCEPTS/CONSTRAINTS FROM THE BEILSTEIN BOZEN SYMPOSIUM ON FUNCTIONAL NANOSCIENCE

The research presented at the Symposium contained a number of common themes, threads, constraints, and methodologies. We have summarized the presentations in terms of these general attributes so that the possible interconnections and interdependencies are drawn out.

Learning from, and instructing biology

Biology provides vivid proof that self-assembling functional nanodevices of immense sophistication and complexity are possible. We are surrounded by (and indeed composed of) a myriad of examples. As well as providing extremely compelling proof of concept, understanding how biological functional nanomachines operate provides a fast track to developing synthetic analogues. These may not have the current constraints of biology in requiring a controlled environment, physiological conditions, and an aqueous environment (extremophiles notwithstanding). As well as providing extremely useful mechanistic ideas, the biological machines themselves can be adapted to perform useful functions. The elegant experiments described by Firman show how DNA helicases can be used to construct functional nanomachines with very useful properties. Biology shows us what is possible if we are clever enough to understand it

We have been tackling difficult problems related to biological nanomachines, wittingly or unwittingly, for a very long time. Historically, this has largely involved disrupting biological processes for medical purposes. Examples include the modulation of DNA replicative machinery by antimicrobial [6, 7] or anticancer drugs (Figure 4), inhibition of enzymes, and artificial activation or inhibition of receptors.

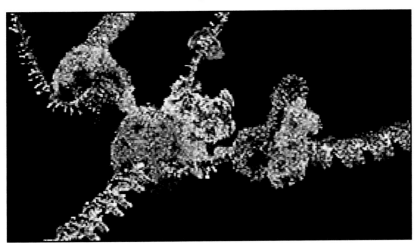

Figure 4. DNA polymerase molecular machine (DNA Replisome image by Drew Berry, the Walter and Eliza Hall Institute).

The stages of maturity in understanding controlling and emulating biological nanomachines may be summarized as follows:

- disrupting/destroying biological nanomachines (drug discovery, > 100 years)
- understanding biological nanomachines (often to learn better how to disrupt them, molecular biology and related sciences, > 50 years)
- emulating/mimicking biological nanomachines (synthetic biology, tissue engineering, present and future of functional nanoscience)
- improving/reprogramming/repurposing biological nanomachines (molecular biology, cell biology (stem cells), functional nanoscience

Disrupting biological nanomachines – drug development

Nanoscience has an important role to play in increasing efficiency of drug design and delivery

Nano approaches to drug targets

Rational drug design often requires detailed knowledge of the 3D structures of relevant molecular targets, usually receptors or enzymes. Although techniques for generating this information using X-ray crystallography has improved greatly, the rate determining step is still the crystallization of the target protein. This is particularly problematic for membrane bound proteins like many of the important receptor targets. Several of the speakers at the Bozen Symposium addressed these and related issues.

Paul Weiss (UCLA, Los Angeles, CA, U.S.A.) described a general technique for immobilizing neurotransmitters on surfaces, and using them to capture membrane receptors for identification, mapping, and structural studies. This may ultimately allow single molecule structural analysis, with no crystallization required [8]. The use of DNA type 1 restriction enzymes and DNA helicases for detecting drug-target, and drug-drug interactions, was described by Firman. The detector he described used the ability of helicase to unwind dsDNA, causing bead movement and a signal. Small molecule inhibitors stopped the helicase from functioning, generating no or low signal. This method may have single molecule sensitivity. Seeman made novel use of DNA origami to generate self-assembled crystal arrays for proteins. These employed a tensegrity triangle [9], and proof of concept was shown by organizing gold nanoparticles into a precisely constructed arrays. He proposed that such arrays might be used as frameworks to solve the protein crystallization problem. Kjems also described how DNA origami might be used to tackle the protein crystallization problem. He demonstrated how DNA arrays or objects could be generated by precisely controlled self-assembly, and how points on DNA objects can be addressed with proteins to make patterns on surface [10]. Shih used DNA origami to generate nanofibres as an NMR alignment medium for solving solution phase structures of membrane-bound proteins. The aligned DNA nanofibres generate a slight alignment bias of proteins allowing structure to be determined [11].

Nano approaches to drug delivery

Presentations by Bozen Symposium speakers described a diverse range of drug delivery methods that aim to approach the "ideal" drug delivery nanoparticle (Figure 5). Jonathan Posner (Arizona State University, Tempe, AZ, U.S.A.) described the properties of motile hybrid Au/Pt nanorods driven by a peroxide reaction. These nano-objects can carry payloads thirty times larger than the nanoparticle, and have potential applications in wound healing, and drug delivery [12]. Kjems' work on DNA origami produced programmable, selective DNA boxes for drug delivery. These have an openable box with a lock closed by hybridization and opened by competing off the DNA oligonucleotide using a free complementary strand that is longer than the lock oligo. The DNA box delivery system can get into cells and appear to be stable in the intracellular environment, but they cannot yet be opened in cells on demand [10]. Tim Clark (University of Erlangen-Nürnberg, Germany) described how theory could lead experiment in design of persistent micelles for drug delivery [13], while Peter

Seeberger (Max Planck-Institute for Colloids and Interfaces, Berlin, Germany) showed examples of carbohydrate targeted nanoparticles. He exploited carbohydrate recognition to allow nanoparticles to be recognized by specific cells and internalized [14]. He has shown proof of concept of the technique applied to siRNA or antisense delivery. Carbohydrate treated quantum dot experiments in live rats showed very good delivery to liver, no obvious toxicity, and the particles were cleared in a few hours.

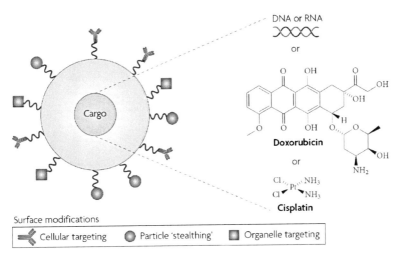

Figure 5. Ideal drug delivery nanoparticle. Petros and de Simone [15] use with permission.

Nano approaches to diagnostics

Functional nanoscience has the potential to revolutionize medical diagnostics. Several novel engineered nanodevices described by the speakers may lead to single molecule detection systems. Firman described the coupling of restriction endonucleases to magnetic beads to generate nanodetectors for quantifying ligand-receptor interactions in an analytical device. Christoph Gerber (University of Basel, Switzerland) described how micro-fabricated silicon cantilever array technology could be used to detect drugs, antibodies, bacteria, and to analyze breath for disease diagnosis. He also illustrated how nanomechanical holography can image live cells and track nanoparticles in them.

The cantilever sensor arrays are based on an AFM without tip, where inkjet technology is used to functionalize cantilever directly. The methods can be used to monitor transcription factors, transcription, and protein synthesis, and can be applied to proteomics by using antibody/antigen interactions. He also described the fast detection of microorganisms by cantilever arrays. These can easily detect resistant versus wild-type microbes (*e.g.* MRSA) where resistance is due to deletion of a *single* hydrogen bond. Seeberger discussed the manufacture of fast, cheap, carbohydrate-based nanodetectors. These exploit automated carbohydrate synthesis, and printing technology applied to the rapid manufacture of glycomic chips [16]. He used nanotechnology to detect pathogenic bacteria quickly and cheaply using surface recognition and a fluorescent polymer. He also described using supramolecular chemistry (cyclodextrans on dendrimers) to make easy cheap detectors for, for example, bacteria and single molecules.

Understanding biological nanomachines

Are living things irreducibly complex?

There is a role for complementary reductionist and complexity-based approaches to understanding properties of biological systems.

Linear scaling quantum mechanical, and fine and coarse-grained MD modelling were described by Clark as useful techniques for understanding properties of biological systems. They provide an appropriate level of scale for modelling biological systems that is coupled to the complexity of the system and the amount of information available from experiments. Petra Schwille (TU Dresden, Germany) discussed experimental models of biological systems such as cells and cell division processes. She is constructing coarse-grained models of cells and genes. This work was informed by single molecule experiments in live organisms to deduce how nanoscale objects in biology self-organize [17]. The aim was to reproduce cellular processes in a simpler way than the cell uses. She adopted a bottom up approach to cell division, starting with a simple cell-like membrane, adding ion channels and skeleton and ATP. Sylvia Speller (Radboud University Nijmegen, The Netherlands) studied real biological systems (magnetoreception and bionavigation in fish) to understand how information is transduced to the nervous systems. Using AFM to study bio-magnetite within the nasal epithelium of salmon, she showed how the observed structures are compatible not only with mechanosensitive gating, but also with voltage gating of ion channels.

What is the basis for the incredibly low error rate with which biological machines operate?

This robust behaviour is a function of high fidelity operations, and editing of errors. In some cases (*e.g.* microtubule assembly) the robustness and self-editing are a function of self-organization, dynamic instability or equilibrium. We need to understand these processes well enough to exploit them in artificial nanomachines. Fraser Stoddart (NWU, Evanston, IL, U.S.A.) tackled this problem by designing "intelligent interactions" into his functional nanodevices [18]. Non-covalent intramolecular interactions can generate precise robust structures, such as catenanes and rotaxanes that can be used to build functional molecular devices. Weiss also worked with precise chemical systems, and attempted to use defects to advantage. Soft lithography developed by Whitesides' group, led to microdisplacement printing that uses adamantanethiols as a stamp "resist" easily displaced by stronger binding alkanethiols or amide substituted versions [19].

Emulating or mimicking biological nano objects

Biological nanomachines have both restrictions and advantages compared to synthetic versions that functional nanoscientists may construct. Primarily, the disadvantages are the need for physiological temperatures, neutral pH, and an aqueous environment. Biological nanomachines can, however, exploit readily available energy currency (ATP) and tap into sophisticated waste disposal processes in organisms. Biological nanomachines are often self-organized, autonomous, and robust. Artificial functional nanodevices, particularly those that move, will require a source of energy. This is a challenging problem, given the extremely small scale of the devices, but experiments described by Paul Weiss, Jonathan Posner, Petr Kràl (University of Illinois, Chicago, IL, U.S.A.), and William Shih show how this may be achieved by exploiting resources in the environment. Peter Dimroth's (ETH Zürich, Switzerland) presentation showed how we might learn from biological motors, ATP synthase being the archetypal example. We may be able to construct artificial nanodevices that exploit ATP and also use the body's waste disposal processes to remove the waste products of their operation.

Functional nanodevices designed for non-biological applications are potentially more problematic. Mimics would be required to operate in a wider range of environments, but need the energy and waste issues resolved. Kràl described how other types of nanopropulsion devices such as propellers, motors, and wheels can be modelled and designed by molecular dynamics calculations (Figure 6) [20]. He also discussed motors driven by electron tunnelling. Nanowheels can be driven by oblique light that generates charges in illuminated regions that drive wheel across water surface. Molecular graphene paddles can also pump liquids.

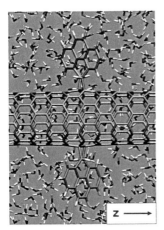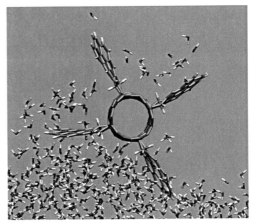

Figure 6. The bulk (left) and surface (right) water propellers that pump water along the tube (z) axis and orthogonal to it, respectively. Both systems are based on carbon nanotubes with covalently attached aromatic blades [20].

However, Kerstin Koch (Rhine-Waal University, Kleve, Germany) showed that nano objects generated by plants have simpler but useful functions that can be emulated. Her studies of the superhydrophobic, self-cleaning surfaces of leaves of plants such as the Lotus, largely due to the nanostructure of the leaf surface, have led to synthetic materials with very useful, self-cleaning properties that can function in diverse environments.

Improving, reprogramming, repurposing biology

Although not specifically covered by the Bozen Symposium presentations, the new research fields of synthetic biology, stem cell science, in addition to functional nanoscience, are beginning to adapt existing biological machines. Reprogramming somatic cells and stem cells to generate induced pluripotent stem cells (iPS cells) are high-level examples of manipulation of very complex biological systems. Directed evolution of enzymes to gain optimized or new functions, or to allow enzymes to operate in non-physiological environments, and the design and delivery of siRNA are additional example. Functional nano-objects that rely on adapting biological systems include nanoactuators based on restriction endonucleases described by Firman, and the exploitation of the properties of DNA to build novel nanostructures discussed by Seeman, Kjems, and Shih.

SELF-ORGANIZATION AND SELF-ASSEMBLY

The terms self-assembly and self-organization are used interchangeably and differently in different scientific disciplines. A recent publication has attempted to provide clear definitions for these terms that are generally applicable [21]. Self-assembly is a non-dissipative structural order on a macroscopic level, because of collective interactions between multiple (usually microscopic) components that do not change their character upon integration into the self-assembled structure. This process is spontaneous because the energy of unassembled components is higher than the self-assembled structure, which is in static equilibrium, persisting without the need for energy input. In contrast, self-organization is a dissipative non-equilibrium order at macroscopic levels, because of collective, nonlinear interactions between multiple microscopic components. This order is induced by interplay between intrinsic and extrinsic factors, and decays upon removal of the energy source. In this context, microscopic and macroscopic are relative. Both forms of programmable pattern formation are useful, but chemists tend to be more familiar with self-assembled objects that essentially form spontaneously using specific interactions between the components that are 'programmed into' the structures. The most mature and robust example of this is the DNA origami discussed in the presentations by Seeman and Shih, although the bottom up work described by most presenters also involved self-assembly. Weiss and Shih emphasized robustness, and precise placement in self-assembled nano objects in their presentations. Stoddard, Seeman, Kjems, and Shih discussed reliable, deterministic rules that result from programmed interactions in self-assembled objects. Quite complex self-assembled nanodevices have been designed, as the photonic device in Figure 7 illustrates [22, 23].

Self-organization has not yet been exploited significantly in functional nanoscience, but examples in nature show that is can be very useful. For example, the dynamic assembly of microtubules from tubulin is partly spontaneous but partly driven by the energy released when GTP is hydrolyzed by tubulin, which is an enzyme as well as a structural component [24]. One paper at the Bozen Symposium focused specifically on self-organization. Schwille described her experiments aimed to emulate some aspects of self-organization in cells. Her work exploits nonlinear dynamical systems reaction diffusion phenomena in biological systems.

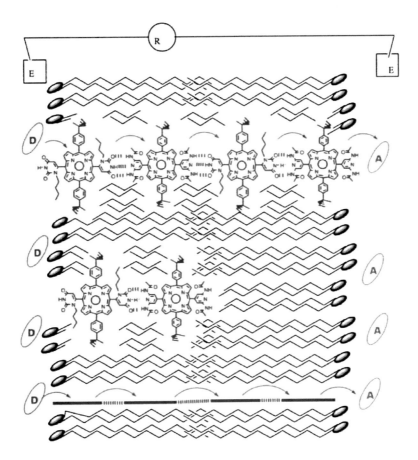

Figure 7. Schematic of one of four self-assembling porphyrin systems self-organized into bilayers to form a functional, self-assembled photonic device [22]. Used with permission.

Recent publications have described a new way to assemble nanostructures by exploiting dynamic instability and incorporating self-healing/editing [25]. Bouchard *et al.* show how to harnessing microtubule dynamic instability for nanostructure assembly. In nature intracellular molecular machines synthesize molecules, dissemble others, transport materials, and transform energy into different forms. Bouchard *et al.* proposed emulating these molecular machines to build nanostructures. Biological nanomachines work in a stochastic, noisy fashion. As described above, microtubules switch randomly between growing and shrinking in a process known as dynamic instability, but the final outcome is a very robust construction of microtubule bundles. A related example is given by kinesin movement along microtubules, which is randomly interrupted by these motor proteins falling off the microtubule. The error recovery afforded by dynamic instability allows these motor proteins to perform their functions autonomously and reliably. Bouchard *et al.* suggested gaining control over these highly dynamic, stochastic processes by eliminating some of them [25]. They show how the natural dynamic instability of microtubules can be exploited to build nanostructures, and described strategies for ensuring that "unreliable" stochastic processes yield a robust outcome. They made extensive use of simulation to understand how to harness dynamic instability, an important component of functional nanoscience discussed in the next section.

Design, computation, simulation and modelling

Most papers presented at the Bozen Symposium stressed the valuable contribution that computational modelling and simulation makes to functional nanoscience. The ability to model the properties of nano objects by molecular dynamics methods was illustrated by Clark (micelles) and Kràl (graphene propellers). Deterministic computational design is tractable, useful, and reliable at least for DNA as illustrated by Seeman, Kjems, and Shih. Stoddart emphasized the importance of "making, modelling, measuring", and Posner successfully modelled nanorod behaviour using simple scaling rules. However, Clark identified important caveats on molecular dynamics calculations for micelle self-assembly. Dimroth showed how modelling can be used to unravel the mechanisms of biological machines like ATP synthase. Athel Cornish-Bowden (CNRS-BIP, Marseilles, France) describes some elegant mathematical modelling to emulate or improve biology, and Michael Huth (University of Frankfurt, Germany) using the Hubbard model to predict the properties of granular metals.

STRENGTHS OF HYBRID TOP DOWN / BOTTOM UP APPROACHES

Limitations of conventional top-down techniques like photolithography and scanning beam (or maskless) lithography include high capital and operating costs, difficulty in accessing these facilities. However, some of these can be overcome by moulding, stamping, and templating methods, and other often simple and elegant methods. An example of the latter is the manufacture of aligned and dimensionally controlled gold nanowires using ice described in Thomas Schimmel's (Karlsruhe Institute of Technology, Germany) presentation.

It was clear from the presentations at the symposium that both top down and bottom up methods of constructing functional nanodevices are capable of yielding very useful and elegant devices. However, the strengths obtained by incorporating techniques and concepts from both approaches came out strongly. Each approach has strengths and weaknesses and these can often be reduce or eliminated by using hybrid approaches.

THE FUTURE

While extrapolation is dangerous and many incorrect projections often result, we have nonetheless attempted to predict where functional nanoscience may lead in the short, medium and long time frames. These extrapolations are based on presentations of the speakers at the Bozen Symposium, discussions with these scientists, and other input provided by nanoscientists not present at the symposium. The predictions will clearly be less accurate as the time frame extends out from 2010.

Extrapolations (2 – 5 years)

Most presenters agree that there will be an increasing move from passive to active nanomachines, for example, nanomachines used as pumps and valves in microfluidics (Gerber). Development of low energy, environmentally benign nanomaterials and manufacturing processes was identified as important drivers by Cornish-Bowden and Schimmel. Coupled to this will be community concern about the safety of nanoparticles and nanodevices, and methods for predicting the impact of nano-objects on cells will begin to appear. Improved methods to control defects will continue to develop, aided by increasingly powerful design, modelling, simulation, and informatics tools. Top down and bottom up fabrication methods will also combined more often in new devices, exploiting the advantages of both techniques (Stoddart). Cantilever devices will grow into larger device arrays and will gain additional functions and speed increases. AFM cantilever devices will begin to be developed for personalized medicine (Gerber). Single molecule structure determination methods, single molecule biosensors (Firman), and single atom quantum switches (Schimmel) will become more robust and begin to be developed into practical devices. Proof of concept for an artificial cell that divides symmetrically will be reported (Schwille), and better modelling of EBID processes, and new metals will be developed (Hans Mulders, FEI Electron Optics, Eindhoven, The Netherlands).

Predictions (5 – 10 years)

Nanomaterials and devices will become increasing complexity and multifunctional. 'Smart materials' such as structural materials with embedded sensors and distributed intelligence will appear, and passive and active nanomaterials will be incorporated into apparel for health monitoring, and sensor networks (Drummond). Faster, larger nanoswitch arrays will be developed (Stoddard, Schimmel) and direct experimental demonstration of a Turing machine may be achieved (Stoddard). The first proof of concept demonstrations of self-organized nanomachines may also be achieved (Winkler). Reliable, targeted drug, vaccine, DNA/RNA delivery systems, with multiple payloads will be demonstrated (multiple, Caruso), and oral delivery of drugs using DNA boxes (Kjems). Very significant progress will be made on solution of the protein crystallization problem using precisely defined nanoarrays decorated with proteins (Seeman), and improved NMR protein structures will be obtained using nanoalignment (Shih). In this time frame, cryoEM images will yield atomic details (Shih), lithographic features as small as to 1 nm will be routinely generated (Huth), and precise control of functionality in carboranes will be possible (Weiss). The glycome will be understood sufficiently to allow precise and reliable targeting of nanoparticles and drugs to specific organs (Seeberger)

Dreams (10 – 20 years)

The increasing level of knowledge about biological systems, and improved nanoscience tools will facilitate the construction of adaptive, self-optimizing nanodevices. Engineered medical nanomachines will be designed that exploit endogenous ATP as a power source. Multiple energy sources such as miniature fuel cells, methods for tapping biological energy for nanodevices will be developed. Nanofabrication defects will be greatly reduced or eliminated by improved fabrication or dynamic self-healing methods, and effective control of self-organization for use in nanoscience will be achieved. Design and constriction of complex, diverse self-assembling nano objects will become routine, and nano-enabled diagnostic devices make personalized medicine a reality. Nano methods of target identification and drug delivery will create new paradigm for successful treatments of diseases. The first commercially useful self-assembled electronic devices will be developed and the estimated market for nanoscience will be US$1 trillion [26].

ACKNOWLEDGEMENTS

The author is very grateful for the helpful insight of the Beilstein Bozen speakers, Calum Drummond (CSIRO Materials Science and Engineering chief and Federation Fellow), Frank Caruso (Melbourne University and Federation fellow), and Richard Williams (CSIRO Materials Science and Engineering). The invitation to attend the symposium and present, and financial support from the Beilstein-Institut is gratefully acknowledged.

REFERENCES

[1] Feynman, R.P. (1960) There's plenty of room at the bottom. *Eng. Sci.* **23**:22 – 36.

[2] Hicks, M.G., Kettner, C. *Proceedings of the International Beilstein Symposium on Systems Chemistry, Bozen, Italy, May 26 - 30, 2008.* Logos-Verlag, Berlin, 2009.

[3]   Hirama, M. (2005) Total synthesis of ciguatoxin CTX3C: a venture into the problems of ciguatera seafood poisoning. *Chem. Rec.* **5**:240 – 50.
doi: 10.1002/tcr.20049.

[4]   Whitesides, G.M., Lipomi, D.,J. (2009) Soft nanotechnology: "structure" vs. "function". *Faraday Discuss.* **143**:373 – 84.
doi: 10.1039/b917540g.

[5]   Nanotechnologies, T. P. o. E. Nanotechnology Consumer Products inventory. http://www.nanotechproject.org/inventories/consumer/

[6]   Wijffels, G., Johnson, W.M., Oakley, A.J., Turner. K., Epa, V.C., Briscoe, S.J., Polley, M., Liepa, A.J., Hofmann, A., Buchardt, J., Christensen, C., Prosselkov, P., Dalrymple, B.P., Alewood, P.F., Jennings, P.A., Dixon, N.E., Winkler, D.A. (2011) Binding Inhibitors of the Bacterial Sliding Clamp by Design. *J. Med. Chem.*, Just Accepted Manuscript. Publication Date (Web): May 23, 2011.
doi: 10.1021/jm2004333.

[7]   Wijffels, G., Dalrymple, B.P., Prosselkov, P., Kongsuwan, K., Epa, V.C., Lilley, P.E., Jergic, S., Buchardt, J., Brown, S.E., Alewood, P.F., Jennings, P.A., Dixon, N.E. (2004) Inhibition of protein interactions with the beta(2) sliding clamp of *Escherichia coli* DNA polymerase III by peptides from beta(2)-binding proteins. *Biochemistry* **43**:5661 – 5671.
doi: 10.1021/bi036229j.

[8]   Vaish, A., Shuster, M.J., Cheunkar, S., Singh, Y.S., Weiss, P.S., Andrews, A.M. (2010) Native Serotonin Membrane Receptors Recognize 5-Hydroxytryptophan-Functionalized Substrates: Enabling Small-Molecule Recognition. *ACS Chem. Neurosci.* **1**:495 – 504.
doi: 10.1021/cn1000205.

[9]   Zheng, J.P., Birktoft, J.J., Chen, Y., Wang, T., Sha, R.J., Constantinou, P.E., Ginell, S.L., Mao, C.D., Seeman, N.C. (2009) From molecular to macroscopic via the rational design of a self-assembled 3D DNA crystal. *Nature* **461**:74 – 77.
doi: 10.1038/nature08274.

[10]   Andersen, E.S., Dong, M., Nielsen, M.M., Jahn, K., Subramani, R., Mamdouh, W., Golas, M.M., Sander, B., Stark, H., Oliveira, C.L.P., Pedersen, J.S., Birkedal, V., Besenbacher, F., Gothelf, K.V., Kjems, J. (2009) Self-assembly of a nanoscale DNA box with a controllable lid. *Nature* **459**:73-U75.
doi: 10.1038/nature07971.

[11] Douglas, S.M., Chou, J.J., Shih, W.M. (2007) DNA-nanotube-induced alignment of membrane proteins for NMR structure determination. *Proc. Nat. Acad. Sci. U.S.A* **104**:6644–6648.
doi: 10.1073/pnas.0700930104.

[12] Burdick, J., Laocharoensuk, R., Wheat, P.M., Posner, J.D., Wang, J. (2008) Synthetic nanomotors in microchannel networks: Directional microchip motion and controlled manipulation of cargo. *J. Am. Chem. Soc.* **130**:8164-+.
doi: 10.1021/ja803529u.

[13] Jager, C.M., Hirsch, A., Schade, B., Bottcher, C., Clark, T. (2009) Counterions Control the Self-Assembly of Structurally Persistent Micelles: Theoretical Prediction and Experimental Observation of Stabilization by Sodium Ions. *Chem. - Eur. J.* **15**:8586–8592.
doi: 10.1002/chem.200900885.

[14] Lepenies, B., Yin, J.A., Seeberger, P.H. (2010) Applications of synthetic carbohydrates to chemical biology. *Curr. Opin. Chem. Biol.* **14**:404–411.
doi: 10.1016/j.cbpa.2010.02.016.

[15] Petros, R.A., DeSimone, J.,M. (2010) Strategies in the design of nanoparticles for therapeutic applications. *Nat. Rev. Drug Discovery* **9**:615–627.
doi: 10.1038/nrd2591.

[16] Seeberger, P.H., Werz, D.B. (2007) Synthesis and medical applications of oligosaccharides. *Nature* **446**:1046–1051.
doi: 10.1038/nature05819.

[17] Loose, M., Fischer-Friedrich, E., Ries, J., Kruse, K., Schwille, P. (2008) Spatial regulators for bacterial cell division self-organize into surface waves *in vitro*. *Science* **320**:789–792.
doi: 10.1126/science.1154413.

[18] Haussmann, P.C., Stoddart, J.F. (2009) Synthesizing Interlocked Molecules Dynamically. *Chem. Rec.* **9**:136–154.
doi: 10.1002/tcr.20173.

[19] Saavedra, H.M., Thompson, C.M., Hohman, J.N., Crespi, V.H., Weiss, P.S. (2009) Reversible Lability by in Situ Reaction of Self-Assembled Monolayers. *J. Am. Chem. Soc.* **131**:2252–2259.
doi: 10.1021/ja807648g.

[20] Wang, B.Y., Kral, P. (2007) Chemically tunable nanoscale propellers of liquids. *Phys. Rev. Lett.* **98**:266102.
doi: 10.1103/PhysRevLett.98.266102.

[21]   Halley, J.D., Winkler, D.A. (2008) Consistent Concepts of Self-organization and Self-assembly. *Complexity* **14**:10 – 17.
        doi: 10.1002/cplx.20235.

[22]   Drain, C.M. (2002) Self-organization of self-assembled photonic materials into functional devices: Photo-switched conductors. *Proc. Nat. Acad. Sci U.S.A.* **99**:5178 – 5182.
        doi: 10.1073/pnas.062635099.

[23]   Milic, T.N., Chi, N., Yablon, D.G., Flynn, G.W., Batteas, J.D., Drain, C.M. (2002) Controlled hierarchical self-assembly and deposition of nanoscale photonic materials. *Angew. Chem., Int. Ed.* **41**:2117 – 2119.
        doi: 10.1002/1521-3773(20020617)41:12<2117::AID-ANIE2117>3.0.CO;2-2.

[24]   Halley, J.D., Winkler, D.A. (2006) Classification of self-organization and emergence in chemical and biological systems. *Aust. J. Chem.* **59**:849 – 853.
        doi: 10.1071/CH06191.

[25]   Bouchard, A.M., Warrender, C.E., Osbourn, G.C. (2006) Harnessing microtubule dynamic instability for nanostructure assembly. *Phys. Rev. E Stat. Nonlin. Soft Matter Phys.* **74**(4).
        doi: 10.1103/PhysRevE.74.041902.

[26]   Sahoo, S.K., Parveen, S., Panda, J.J. (2007) The present and future of nanotechnology in human health care. *Nanomedicine* **3**:20 – 31.
        doi: 10.1016/j.nano.2006.11.008.

Functional Nanoscience
May 17th – 21st, 2010, Bozen, Italy

 Beilstein-Institut

BIOGRAPHIES

Timothy Clark

was born (1949) in southern England and studied chemistry at the University of Kent at Canterbury, where he was awarded a first class honours Bachelor of Science in 1970. He obtained his Ph.D. from the Queen's University Belfast in 1973 after working on the thermochemistry and solid-phase properties of adamantane and diamantane derivatives. After two years as an Imperial Chemical Industries Postdoctoral Fellow in Belfast, he moved in 1975 to Princeton University as a NATO Postdoctoral Fellow working for Paul Schleyer. He then followed Schleyer to the Institut für Organische Chemie of the Universität Erlangen-Nürnberg in 1976, where he subsequently habilitated. He is currently Technical Director of the Computer-Chemie-Centrum in Erlangen. His research areas include the development and application of quantum mechanical and classical simulation methods in inorganic, organic and biological chemistry, electron-transfer theory and the simulation of reaction mechanisms, especially enzymatic, and the mechanisms of signal-transduction processes. He is the developer of the semiempirical molecular orbital program VAMP, which is marketed as a compute-engine in Accelrys' Materials Studio Modeling package and of the surface-property package ParaSurf, which is marketed by Cepos InSilico Ltd., a UK-based software company of which he is director and CEO. Clark is a member of the Scientific Advisory Board of Boehringer Ingelheim Pharma AG. He is the author of 300 articles in scientific journals and two books and is the founding editor of the *Journal of Molecular Modeling*. He was twice Visiting Professor in the Department of Chemistry of the University of Wisconsin, Madison and was named part time Professor of Computational Chemistry (Centre for Molecular Design, School of Pharmacy and Biomedical Science) of Portsmouth University, UK, in February 2006 in addition to his position in Erlangen. In 2009, he was awarded the Klaus-Wilhelm von der Lieth Medal by the Molecular Graphics and Modelling Society. He is a principal investigator in SFB473 and SFB583 and principal investigator and member of the board of the Excellence Cluster Engineering of New Materials.

Athel Cornish-Bowden

carried out his undergraduate and postgraduate studies at Oxford, with research on pepsin catalysis under the direction of Jeremy R. Knowles leading to his doctorate in 1967. He spent three years as a post-doctoral fellow in the laboratory of Daniel E. Koshland, Jr., at the University of California at Berkeley, where he was principally concerned with subunit interactions in proteins. He was Lecturer in Biochemistry at the University of Birmingham from 1970 until 1985, and then Senior Lecturer until 1987, when he moved to Marseilles as

Directeur de Recherche of the CNRS. Since September 2009 he has been Directeur de Recherche Émérite. Despite starting his career in a department of organic chemistry, virtually all of his research has been in biochemistry, especially enzyme kinetics, a topic on which he has written several books, including *Analysis of Enzyme Kinetic Data* (Oxford University Press, 1995) and *Fundamentals of Enzyme Kinetics* (3rd edition, Portland Press, 2004).

His work on the kinetics of individual enzymes led naturally to an interest in the role of enzymes in metabolic systems, and hence metabolic regulation and control. In recent years he has a major interest in the definition of life and the capacity of organisms construct and maintain themselves without requiring external catalysts. This capacity for metabolic circularity constitutes an essential difference between mechanical devices and organisms. As the central main topic of his research, he has long had an interest in biochemical aspects of evolution, a topic developed in his book *The Pursuit of Perfection* (Oxford University Press, 2004), and the subject of lecture series this year in Belgium and in Chile.

This is his 5th participation in the series of Beilstein Symposia in Bolzano. He is also very active in the Beilstein ESCEC Symposia and the committee STRENDA (Standards for Reporting Enzymology Data) created by the Beilstein Institute.

Peter Dimroth

Professional experience:

1990-   Professor of Microbiology, ETH Zürich, Institute of Microbiology

1980 – 1990 Professor of Physiological Chemistry, Technical University Munich

1971 – 1979 Research Assistant, University of Regensburg

1969 – 1971 Postdoctoral fellow, New York University and Johns Hopkins University

University education:

1965 – 1969 PhD thesis, Max-Planck-Institute for Cell Chemistry, Munich

1960 – 1965 Studies of Chemistry

Awards:

1977   Karl-Winnacker-Prize

Keith Firman

is employed as a Reader in Molecular Biotechnology, in the Biophysics Laboratories, School of Biological Sciences, University of Portsmouth since 2002, where he leads a research group exploring various areas of the use of molecular machines as biosensors and nanoactuators in bionanotechnology. Recently he has coordinated two, consecutive EC-funded projects that have allowed him to demonstrate, at the single molecule level, the capability of a biological motor to 'pull' micron-sized objects over microns distance. This work has recently attracted a further grant from EPSRC under the EUROCORES Scheme NanoSci-E+) to demonstrate the use of single molecule nanoactuators for drug-target investigations. This work has led to the commercial development of the technology in the form of a single-molecule biosensor which can link to existing sensing systems, or be used in novel environments such as toxicity testing or drug discovery.

1996 – 2002   Promoted to Principal Lecturer based on experience as Course Leader for Molecular Biology Degree Pathway and extensive research experience.

1988 – 1996   Senior Lecturer in the School of Biological Sciences of Portsmouth Polytechnic (now University).

1982 – 1988   Medical Research Council post-doctoral research assistantships (two) at University of Newcastle upon Tyne.

1979 – 1982   Part-time PhD at University of Newcastle upon Tyne.

1974 – 1979   Research technician at the Dept of Genetics, Newcastle University, initially working on the isolation of unique plasmid borne restriction and modification systems. During this time he was re-graded from grade 4 to grade 6 technician. This re-grading was based exclusively upon my research capabilities and not upon the acquisition of other duties.

1972 – 1974   University research technician (grade 4). University of Hull – Biochemistry.

Martin G. Hicks

is a member of the board of management of the Beilstein-Institut. He received an honours degree in chemistry from Keele University in 1979. There, he also obtained his PhD in 1983 studying synthetic and theoretical approaches to the photochemistry of pyridotropones under the supervision of Gurnos Jones. He then went to the University of Wuppertal as a post-doctoral fellow, where he carried out research with Walter Thiel on semi-empirical quantum chemical methods. In 1985, Martin joined the computer department of the Beilstein-Institut where he worked on the Beilstein Database project. His subsequent activities involved the development of cheminformatics tools and products in the areas of substructure searching and reaction databases.

Thereafter, he took on various roles for the Beilstein-Institut, including managing director-ships of subsidiary companies and was head of the funding department 2000 – 2007. He joined the board of management in 2002 and his current interests and responsibilities range from organization of Beilstein Symposia with the aim of furthering interdisciplinary com-munication between chemistry and neighbouring scientific areas, to the publishing of Beil-stein Open Access journals such as the Beilstein Journal of Organic Chemistry and the Beilstein Journal of Nanotechnology.

Michael Huth

received his diploma and doctoral degree in physics at the Technical University Darmstadt in 1990 and 1995, respectively. After having spent some time as a research assistant at the Johannes Gutenberg University Mainz he went with a DFG research scholarship to the Physics Department and Materials Research Laboratory of the University of Illinois at Urbana-Champaign. He returned in 1998 to the University of Mainz were he completed his habilitation in experimental physics in 2001. Since 2002 he is professor for experimental condensed matter physics at the Goethe University Frankfurt am Main.

Michael Huth is co-founder of NanoScale Systems GmbH, a small company that develops new sensor concepts based on maskless lithography using focused particle beam induced deposition techniques.

Carsten Kettner

studied biology at the University of Bonn and obtained his diploma at the University of Göttingen. Here, in Prof. Gradmann's group "Molecular Electrobiology" which consisted of people carrying out research in electrophysiology and molecular biology in fruitful coopera-tion, he studied transport characteristics of the yeast plasma membrane using patch clamp techniques. In 1996 he joined the group of Dr. Adam Bertl at the University of Karlsruhe and successfully narrowed the gap between the biochemical and genetic properties, and the biophysical comprehension of the vacuolar proton-translocating ATP-hydrolase. He was awarded his Ph.D for this work in 1999. As a post-doctoral student he continued both the studies on the biophysical properties of the pump and investigated the kinetics and regula-tion of the dominant plasma membrane potassium channel (TOK1). In 2000 he moved to the Beilstein-Institut to represent the biological section of the funding department. Here, he is responsible for the organization of the Beilstein symposia, research (proposals) and publica-tion of the proceedings of the symposia. Since 2004 he coordinates the work of the STRENDA commission and promotes along with the commissioners the proposed standards of reporting enzyme data. In 2007 he became involved in the development of a program for the establishment of Beilstein Endowed Chairs for Chemical Sciences and related sciences. At the same time he started a correspondence course at the Studiengemeinschaft Darmstadt (a certified service provider) where he was awarded his certificate of competence as project manager for his studies and thesis. In 2009, he took part in the establishment of a large-scale funding project in the nanoscience area called NanoBiC.

Kerstin Koch

Scientific education:

- Biology (Diploma) in 1992 and PhD in 1998 at the University of Bonn.
- 2001 – 2006 Assistant Professor at Nees-Institute for Biodiversity of Plants (Bonn University) and habilitation in 2006.
- 2007 Associate Professor at Bonn University.
- 2008 visiting Professor at the Ohio State University (USA), Nanoprobe Lab. for Bio- & Nanotechnology and Biomimetics (NLB2).
- Since Sept. 2009 Prof. for Biology and Nanobiotechnology at the University of Applied Sciences, Rhine-Waal, Kleve.

Main fields of research:

function and origin of micro- and nanostructures of biological surfaces; boundary layer interactions, e.g. wetting of surfaces, molecular self-assembly on surfaces, molecular architectur and crystallinity of plant waxes, development of biomimetic surfaces, e.g. with replica techniques.

Petr Král

2004-Present	Assistant Professor, University of Illinois at Chicago, Department of Chemistry

Education

1986	BS and MS, Czech Technical University, Prague
1987	Military service
1987 – 1988	RA, Department of Magnetism, Czech Academy of Sciences (CAS) and Department of Quantum Optics, Palack'y University
1989 – 1990	RA, Department of Condensed Matter Theory, CAS
1995	PhD, Department of Condensed Matter Theory, CAS; Thesis: "Quantum Theory of Linear Electron Transport in quasi– 1D Semiconductor Structures"

Professional Experience

1996 – 1997	University of Nottingham, NATO/Royal Society Fellowship
1997 – 1999	University of Toronto, Postdoctoral Fellow
1999 – 2001	Weizmann Institute of Science, Postdoctoral Fellow
2001 – 2004	Weizmann Institute of Science and Harvard University, Visiting Scientist
2003 – 2004	University of British Columbia, Research Associate
2004	University of Illinois at Chicago, Assistant Professor of Chemistry

Jonathan D. Posner

is an assistant professor at Arizona State University in the faculties of Mechanical Engineering and Chemical Engineering as well as adjunct faculty in the Consortium for Science, Policy, & Outcomes (CSPO). He is the director of the ASU Micro/Nanofluidics Lab and the PI of eight ongoing projects. Dr. Posner earned his Ph.D. (2001) degree in Mechanical Engineering at the University of California, Irvine. He spent 18 months as a fellow at the von Karman Institute for Fluid Mechanics in Rhode Saint Genèse, Belgium and two years as a postdoctoral fellow at the Stanford University. His interests include electrokinetics, self-assembly of colloids, and the physics of nanoparticles at interfaces. At CSPO, Posner has interest in the social implications of technology, role of science in policy and regulation, as well as ethics education. Dr. Posner was honored with a 2008 NSF CAREER award for his work on the physics of self-assembly of nanoparticles at fluid-solid and fluid-fluid interfaces. He has also been recognized for his Excellence in Experimental Research by the von Karman Institute for Fluid Dynamics.

Thomas Schimmel

did his PhD in Physics at the University of Bayreuth, Germany. One year after being nominated as Post-doctoral Research Fellow at BASF company, he returned to the University of Bayreuth, starting a group on scanning probe microscopy at the University of Bayreuth. Schimmel received offers for professorial positions at the Johannes Kepler University of Linz, Austria, the Ludwig Maximilians University, Munich, and the University of Karlsruhe.

Since 1996, he is Professor and joint Institute Director of the Institute of Applied Physics at the University of Karlsruhe. Furthermore, he is co-founder of the Institute of Nanotechnology at the Karlsruhe Institute of Technology (KIT) and Head of Department at this institute since 1998. In 2002, Schimmel was nominated Foreign Advisory Professor at Shanghai Jiaotong University. Schimmel is also initiator and Director of the Research Network of Excellence on Functional Nanostructures in Baden-Württemberg (since 2002).

Schimmel's research focuses on nanoscience and nanotechnology. His scientific prizes and awards include the Philip Morris Research Prize, the Emil Warburg Award and the Research Prize of the State of Baden-Württemberg.

Peter H. Seeberger

received his Vordiplom in 1989 from the Universität Erlangen-Nürnberg, where he studied chemistry as a Bavarian government fellow. In 1990 he moved as a Fulbright scholar to the University of Colorado where he earned his Ph.D. in biochemistry under the guidance of Marvin H. Caruthers in 1995. After a postdoctoral fellowship with Samuel J. Danishefsky at the Sloan-Kettering Institute for Cancer Research in New York City he became Assistant Professor at the Massachusetts Institute of Technology in January 1998 and was promoted to Firmenich Associate Professor of Chemistry with tenure in 2002. From June 2003 until January 2009 held the position of Professor for Organic Chemistry at the Swiss Federal Institute of Technology (ETH) in Zurich, Switzerland, where he served as chair of the laboratory in 2008. In 2009 he assumed positions as Director at the Max-Planck Institute for Colloids and Surfaces in Potsdam and Professor at the Free University of Berlin. Since 2003 he serves as an Affiliate Professor at the Sanford-Burnham Institute in La Jolla, CA.

Professor Seeberger's research has been documented in over 220 articles in peer-reviewed journals, two books, fifteen issued patents and patent applications, more than 100 published abstracts and more than 450 invited lectures. Among other awards he received the Arthur C. Cope Young Scholar and Horace B. Isbell Awards from the American Chemical Society (2003), the Otto-Klung Weberbank Prize for Chemistry (2004), the Havinga Medal (2007), the Yoshimasa Hirata Gold Medal (2007), the Körber Prize for European Sciences (2007), the UCB-Ehrlich Award for Excellence in Medicinal Chemistry (2008) and the Claude S. Hudson Award for Carbohydrate Chemistry from the American Chemical Society (2009). In 2007 and 2008 he was selected among "The 100 Most Important Swiss" by the magazine "Schweizer Illustrierte". In 2010 he will receive the Tetrahedron Young Investigator Award for Bioorganic and Medicinal Chemistry.

Peter H. Seeberger is the Editor of the *Journal of Carbohydrate Chemistry* and serves on the editorial advisory boards of seven other journals. He is a founding member of the board of the Tesfa-Ilg "Hope for Africa" Foundation that aims at improving health care in Ethiopia in particular by providing access to malaria vaccines and HIV treatments. He is a consultant and serves on the scientific advisory board of several companies. In 2006 he served as president of the Swiss Academy of Natural Sciences.

The research in Professor Seeberger's laboratory has resulted in two spin-off companies: Ancora Pharmaceuticals (founded in 2002, Medford, U.S.A.) that is currently developing a promising malaria vaccine candidate in late preclinical trials as well as several other therapeutics based on carbohydrates and i2chem (2007, Cambridge, U.S.A.) that is commercializing microreactors for chemical applications.

Paul S. Weiss

is the Director of the California NanoSystems Institute, Fred Kavli Chair in NanoSystems Sciences, and Distinguished Professor of Chemistry and Biochemistry at the University of California, Los Angeles. He received his S.B. and S.M. degrees in chemistry from MIT in 1980 and his Ph.D. in chemistry from the University of California at Berkeley in 1986. He was a post-doctoral member of technical staff at Bell Laboratories from 1986–1988 and a Visiting Scientist at IBM Almaden Research Center from 1988–1989. Before coming to UCLA in 2009, he was a Distinguished Professor of Chemistry and Physics at the Pennsylvania State University, where he began his academic career as an assistant professor in 1989. His interdisciplinary research group includes chemists, physicists, biologists, materials scientists, electrical and mechanical engineers, and computer scientists. Their work focuses on the atomic-scale chemical, physical, optical, mechanical and electronic properties of surfaces and supramolecular assemblies. He and his students have developed new techniques to expand the applicability and chemical specificity of scanning probe microscopies.

They have applied these and other tools to the study of catalysis, self- and directed assembly, physical models of biological systems, and molecular and nano-scale electronics. They work to advance nanofabrication down to ever smaller scales and greater chemical specificity in order to connect, to operate, and to test molecular devices. He has published over 200 papers and patents, and has given over 400 invited and plenary lectures.

Weiss has been awarded a National Science Foundation Presidential Young Investigator Award (1991–1996), the Scanning Microscopy International Presidential Scholarship (1994), the B. F. Goodrich Collegiate Inventors Award (1994), an Alfred P. Sloan Foundation Fellowship (1995–1997), the American Chemical Society Nobel Laureate Signature Award for Graduate Education in Chemistry (1996), a John Simon Guggenheim Memorial Foundation Fellowship (1997), and a National Science Foundation Creativity Award (1997–1999), among others. He was elected a Fellow of: the American Association for the Advancement of Science (2000), the American Physical Society (2002), and the American Vacuum Society (2007). He was also elected a senior member of the IEEE (2009). He received Penn State's University Teaching Award from the Schreyer Honors College (2004), was named one of two Nanofabrication Fellows at Penn State (2005), and won the Alpha Chi Sigma Outstanding Professor Award (2007). He was a Visiting Professor at the University of Washington, Department of Molecular Biotechnology from 1996–1997 and at the Kyoto University, Electronic Science and Engineering Department and Venture Business

Laboratory in 1998 and 2000. Weiss has been a member of the editorial boards of *Review of Scientific Instruments* (2000 – 2005) and other journals, the American Vacuum Society Nanoscience and Technology Division Executive Committee (2001 – 2003) and the U.S. National Committee to the International Union of Pure and Applied Chemistry (2000 – 2005). He has been the Technical Co-chair of the Foundations of Nanoscience Meetings, Thematic Chair of the Spring 2009 American Chemical Society National Meeting, and the Chair of the 2009 International Meeting on Molecular Electronics. He was the Senior Editor of *IEEE Electron Device Letters* for molecular and organic electronics (2005 – 2007), and is the founding Editor-in-Chief of *ACS Nano* (2007-). At *ACS Nano*, he won the Association of American Publishers, Professional Scholarly Publishing PROSE Award for 2008, Best New Journal in Science, Technology, and Medicine.

Dave A. Winkler

studied chemistry, chemical engineering, and physics at the Monash and RMIT Universities in Melbourne. He completed a PhD in microwave spectroscopy and radioastronomy at Monash University in 1980 as a General Motors postgraduate fellow. He then worked as a tutor and senior research fellow, helping establish the computer-aided drug design group at the Victorian College of Pharmacy under Prof. Peter Andrews, in the early 80 s. He subsequently worked as a senior research scientist with the Defence Science and Technology organization in Adelaide for two years, modelling energetic polymer properties. He joined CSIRO Applied Organic Chemistry in Melbourne as a senior research scientist in 1985 where he worked on the design of biologically active small molecules as drugs and agrochemicals. He was awarded Australian Academy of Science traveling fellowships to Toshio Fujita's lab in Kyoto in 1988, and to Graham Richards' research group in Oxford in 1997. He has remained with CSIRO and his worked a common theme of the application of computational methods to understanding and designing novel small molecules with biological activity. During the last decade he has worked with a diverse complex systems science group in Australia and internationally. This aimed to understand how complexity concepts such as emergence, criticality, self-organization and self-assembly can contribute to the modelling, understanding and design of complex chemical and biological systems. In the past five years, his research has evolved to encompass nanotechnology, biomaterials and regenerative medicine. His particular interests in nanotechnology are: predictive nanotoxicology; and modelling self-assembly of nanoparticles and other molecular systems. Dave is a Fellow and past Chairman of the Board of the Royal Australian Chemical Institute, an Adjunct Professor at Monash University, and past President of Asian Federation for Medicinal Chemistry. He has published almost 150 scientific papers, confidential reports, and patents, co-edited three multi-author books. He is a member of the editorial board of journal *ChemMedChem*.

Index of Authors

Index

Index